# Food Emulsion and Foams: Theory and Practice



*Elmer L. Gaden, Jr., series editor*

*Peter J. Wan, volume editor*

*John L. Cavallo, Fouad Z. Saleeb and Michael J. McCarthy, coeditors*

*Maura Mullen, manager publications production*

N.M. Barfod
Rajendra P. Borwankar
Gregory S. Buliga
John L. Cavallo
David L. Chang
J.B. German
Gerald L. Hasenhuettl
Juliana R. Heil
John M. Krochta
N. Krog
J. Lakkis
Robert G. Laughlin
Lloyd A. Lobo
Kathryn L. McCarthy
M.J. McCarthy
Ganesan Narsimhan
David S. Reid
Fouad Z. Saleeb
R. Villota
Darsh T. Wasan

**Inquiries regarding the publication of Symposium Series Volumes should be directed to**

**Dr. Elmer L. Gaden, Jr., series editor,**
**University of Virginia, Department of Chemical Engineering,**
**Thornton Hall, Charlottesville, Virginia 22903-2442. FAX (804) 924-6270.**

AIChE Symposium Series

Number 277 | 1990 | Volume 86

*Published by*

American Institute of Chemical Engineers

345 East 47 Street | New York, New York 10017

American Institute of Chemical Engineers
345 East 47 Street, New York, N.Y. 10017

*AIChE shall not be responsible for statements or opinions advanced in papers or printed in its publications.*

**Library of Congress Cataloging-in-Publication Data**

Food emulsion and foam :/theory and practice / Peter J. Wan, volume editor ; John L. Cavallo, Fouad Z. Saleeb, and Michael J. McCarthy, volume co-editors.

p. cm. — (AIChE symposium series ; no. 277, v. 86, 1990)

Papers presented at the 1989 AIChE Annual Meeting, held in San Francisco, Nov. 5-10, 1989.

Includes index.

ISBN 0-8169-0494-4

1. Emulsions—Congresses. 2. Foam—Congresses. 3. Food—Congresses. I. Wan, Peter J. II. American Institute of Chemical Engineers. Meeting (1989 : San Francisco, Calif.) III. Series: AIChE symposium series ; no. 277.

TP156.E6F67 1990

664—dc20 90-1260

CIP

# FOREWORD

In 1988, we were asked to organize a Food Emulsion Symposium for the AIChE Annual Meeting in San Francisco, November 5–10, 1989. We started with the idea of including all three components, namely, theory, methods and applications, in one symposium. Soon we realized that there was too much to be covered in one session. Finally, the topics were divided into two separate symposia. One dealt with theoretical subjects of food emulsion. The other covered selected practical examples of food emulsions.

The speakers were from both academic and industrial sectors. All the speakers committed to these symposia had a common understanding that there is a need for a food emulsion publication. The feeling grew stronger after the AIChE meeting. At the same time, we discovered that a very good symposium on foam was also presented during the same convention. After several communications with the symposium chairmen and speakers, a proposal to combine food emulsion and foam in one volume was soon approved by the AIChE Symposium Series Editor.

The focus of these symposia is to demonstrate the importance of surface and colloid science in food systems. The topics covered the interfacial properties of emulsifier and protein films, phase science of food emulsifiers, formulation considerations in emulsion systems; oil-in-water (O/W), water-in-oil (W/O) or air-in-liquid (A/L), interfacial rheology, foam rheology, foam stability, and water mobility in foam systems. Design and selection of food emulsifiers, principles of emulsion films and examples of several food emulsions, such as, margarine, low fat spreads, and ice creams were also presented.

The state of the art of methodologies and hardware were presented at the symposium but failed to make this publication. We sincerely hope that this volume provides a better understanding on food emulsions and offers some useful ideas on such complex food systems to the readers. We also wish to express our gratitude to all our speakers, authors and reviewers who made this publication a successful and meaningful task.

Peter J. Wan, *volume editor*
Food and Feed Processing Research
Southern Regional Research Center
USDA-ARS
New Orleans, LA

## CONTENTS

# INTERFACIAL PROPERTIES OF EMULSIFIER/PROTEIN FILMS RELATED TO FOOD EMULSIONS

N. Krog and N.M. Barfod ■ Grindsted Products, 8220 Brabrand, Denmark

In many food emulsions the fat globules are stabilized to prevent coalescence by an adsorbed protein film. When emulsions are aerated (creams, toppings, ice cream mix) the protein film is partially removed by mechanical shear forces during whipping. The increased hydrophobicity of the fat globules causes them to be adsorbed at the air/water interface during aeration. The air cells of the final foam are thus covered by a layer of flocculated fat globules, and fat crystals may be present as well. The stability, firmness and texture of the foam are determined by the interconnecting shells of fat particles that it contains. In the case of ice cream, partial protein desorption from fat globules takes place during the ageing of the mix before freezing. This desorption is enhanced by emulsifiers, which change the interfacial film properties when the emulsion is cooled. The present study shows a correlation between the protein desorption from the fat globule surfaces and the interfacial film properties of oil/water systems containing monoglycerides and proteins.

Many food emulsions are stabilized by proteins, which form a protective layer round the fat globules. In aerated emulsions such as ice cream, frozen desserts, whipped cream and toppings the air cells are stabilized by adsorbed fat globules, fat crystals and proteins. In simple milk foams the air cells are covered with milk casein micelles, which stabilize such foams (1). During the aeration of cream the initial foam formation is also created by the proteins which are present, but on continued whipping fat globules become increasingly adsorbed at the air/serum interface together with fat crystals from mechanically disrupted fat globules. The air cells are thus predominantly stabilized by fat globules or crystals in such foams. Whipped toppings based on high lauric type fats undergo a complete structural change during reconstitution and whipping, resulting in a foam stabilized mainly by fat crystal platelets at the air/serum interface (2).

GRINDSTED PRODUCTS
DK - 8220 Brabrand, Denmark

Paper presented at the
American Institute of Chemical
Engineers, 1989 Annual Meeting
San Francisco, Nov. 5-10, 1989

Food emulsifiers are used to control the destabilization of such emulsions, and their main function is to promote desorption of inter-facially-bound protein, making the fat globules more hydrophobic, increasing their affinity for air, and thus enhancing their adsorption at the air cell interface (3). A correlation between a decrease in interfacial tension and an increase in fat destabilization in the presence of emulsifiers has been described in the literature (4). In order to study the functions of monoglycerides further, interfacial tension measurements of oil/water systems similar to whippable emulsions have been studied at temperatures corresponding to the processing conditions for ice cream in an attempt to explain the role of emulsifiers in whipped emulsions in greater detail.

## THE ADSORPTION OF PROTEINS

Protein molecules may adsorb to oil droplets with several segments of the same molecule in contact with the oil phase and other more hydrophilic segments arranged in the bulk water phase. If the oil contains surface-active lipids, such as monoglycerides, which adsorb more rapidly than proteins, these lipids may provide an energy barrier against protein adsorption, and thereby influence the amount of protein adsorbed (5).

In general, the function of proteins is to give steric stabilization of emulsions against coalescence, and their relative effect depends on their solubility and surface activity. Small globular proteins such as bovine serum albumin and lysozyme are highly soluble but limited with regard to surface activity, whereas large hydrophobic proteins (such as heat-denatured proteins) are sufficiently surface-active but limited with regard to solubility. The best compromise is proteins such as sodium caseinate which are both highly soluble and surface-active.

Casein in milk consists of four types of proteins known as α-s1, α-s2, β, and κ-casein, which are present in the ratio 4:1:4:1. They can be co-precipitated at pH 4.6 and made soluble by forming sodium caseinates. It has been found that α-s1 and β-caseinates are equally adsorbed during homogenization, but on ageing β-casein replaces some but not all surface-bound α-s2 casein (6). β-casein is also found to be the most surface-active and the most hydrophobic of the milk proteins. When using skimmed milk for making emulsions (recombined milk), all the caseins are adsorbed onto the surfaces of oil droplets in the same proportions as in milk (7). Under such conditions caseins are adsorbed as micelles, and it is found that large micelles (up to 300 nm) are adsorbed in preference to small micelles or serum proteins (8).

An important factor for steric stabilization of oil droplets is the amount of protein adsorbed at the surface, the so-called surface load, normally expressed in mg protein per $m^2$ surface area of emulsified oil. The protein load may vary considerably in o/w emulsions, e.g. from 1 $mg/m^2$ (roughly corresponding to a monolayer of extended unfolded polypeptide chains) to several times 10 $mg/m^2$ (9). A surface load of approx. 3 $mg/m^2$ corresponds to a monolayer of globular proteins or partly unfolded protein molecules adsorbed in loops and trains. If the surface load is greater than 5 $mg/m^2$, the protein layer consists of aggregates or multilayers. The surface load varies, depending on the concentration of protein in bulk solution and on the oil surface area formed during emulsification. The proteins of skimmed milk display complete adsorption up to a surface load of about 8 $mg/m^2$, which is a saturation point. Whey proteins and sodium caseinate have a saturation point of about 2.5-3 $mg/m^2$.

In food emulsions such as ice cream mixes,proteins and emulsifiers as well as hydrocolloids (gums) are usually all present, and these components all take part in forming the surface film around the oil droplets. The emulsifiers are usually monoglycerides or blends of monoglycerides and polysorbates in a ratio of about 4:1, and the emulsifiers are more surface-active than milk proteins or gums. Consequently, the emulsifiers will be adsorbed in preference to the o/w interface due to their higher adsorption energy, and thereby influence the adsorption of proteins. A reduction in protein load of as much as 50% has been found when highly polar emulsifiers such as polysorbates are used (4).

## CONTROLLED DESTABILIZATION OF WHIPPABLE EMULSIONS

The structure of aerated emulsions such as ice cream, whipped cream and toppings consists of air cells distributed uniformly in a frozen or liquid continuous phase containing carbohydrates, proteins and fat globules. The texture and stability of the whipped product is related to the air cell structure, which is built up by shells of fat globules or crystals arranged at the interface between air and water.

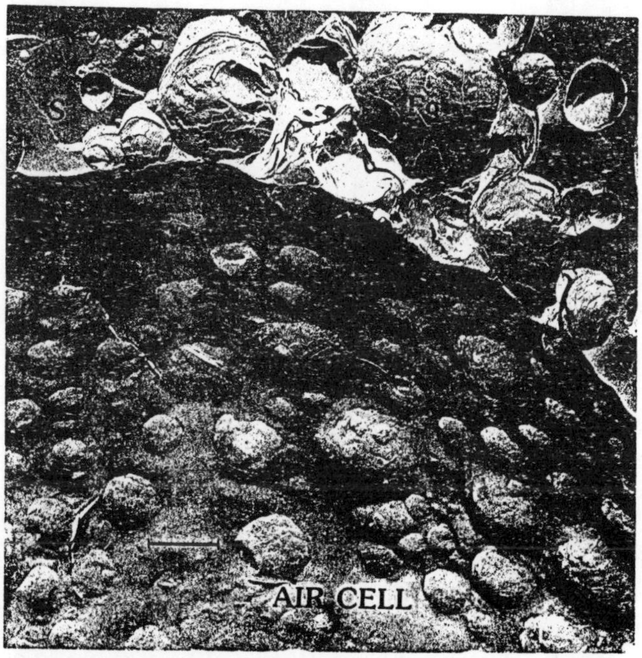


Figure 1. The structure of the air cell/serum interface in a whipped imitation cream by transmission electron microscopy using freeze fracture technique. S = serum phase, Fg = fat globule. Bar = 1 $\mu$m. By courtesy of Dr. W. Buchheim, Kiel.

Figure 1 shows the air cell structure of an aerated, imitation dairy cream demonstrated by electron microscopy using the freeze fracture technique (3). The fat globules are adsorbed at the air-water interface, and protrude into the air cell to varying degrees. The driving force for this process is partial destabilization of the fat globules, from which the protective protein membrane is partly re-moved, making the fat globules more hydrophobic and thus forcing them to be oriented towards the air-water interface. Destabilization involves both a desorption of protein from the fat globule surface and crystallisation of the fat phase, both of which take place during the ageing period of an ice cream mix before freezing. Both these processes are essential for the final volume, texture and stability of ice cream, and both are enhanced by monoglycerides and other surface-active lipids.

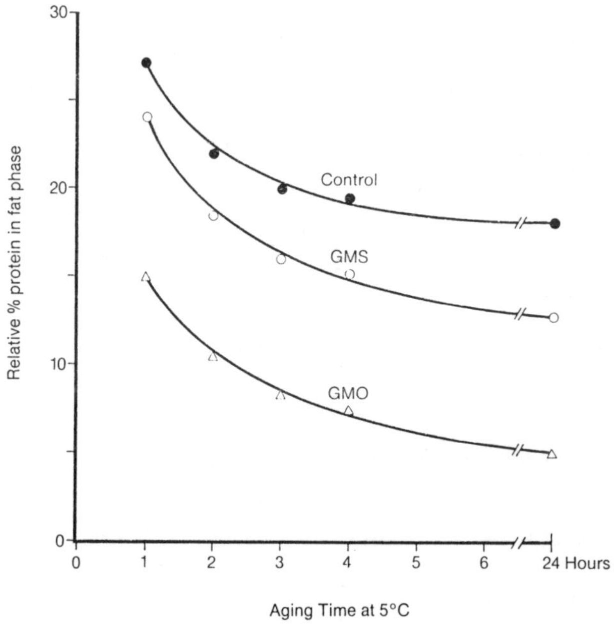


Figure 2. Changes in amount of protein absorbed to fat globules in an ice cream mix during the ageing period at 5°C. The amount of protein bound in the fat phase is calculated relative to the total protein content (4.4%) of the ice cream mix.

The desorption of protein from fat globules in ice cream mix during ageing is demonstrated in Figure 2.

Ice cream emulsions containing 10% fat with or without 0.2% emulsifiers were homogenized at 140 bar at 80°C, cooled to 5°C and aged at 5°C up to 24 hours. The emulsifiers used were distilled, saturated monoglycerides (DIMODAN PV, Grindsted Products, Denmark), referred to as GMS, and distilled, mono-unsaturated monoglycerides, iodine value 55 (DIMODAN OT, Grindsted Products, Denmark) referred to as GMO. Samples of each emulsion were taken at one hour intervals and centrifuged at low speed in order to separate the fat phase from the serum phase without subjecting the fat globules to coalescence. The protein content was then determined in the serum phase using the Kjeldahl method. The amount of protein bound to the fat phase was calculated as the difference between the total protein content of the whole emulsion and the amount of protein in the serum. When the mix was cooled to 5°C, desorption of interfacial protein took place. The addition of emulsifiers enhances the desorption process to varying degrees.

After homogenization of the emulsions, the protein load on the fat globules is lower when emulsifiers are added to the fat phase than when no emulsifiers are present. The actual protein load appears to decrease strongly at the initial stage of ageing at 5°C, especially with unsaturated monoglycerides (GMO) added. After four hours' ageing protein desorption levels off and only a slight decrease in protein load is found in the period between 4 hours and up to 24 hours' ageing time. The final protein load of the fat globules is determined by the type of emulsifier added. The emulsifiers which give a higher surface pressure (i.e. lower interfacial tension) than proteins at interfacial films will have a displacing effect on the proteins.

INTERFACIAL TENSION MEASUREMENTS

Oil/water systems containing milk proteins and emulsifiers were studied using the Krüss Tensiometer Model K10. The temperature of the o/w system was controlled by a thermostat which increased or decreased the temperature at a given rate. The interfacial tension was measured by the Wilhelmy plate method.

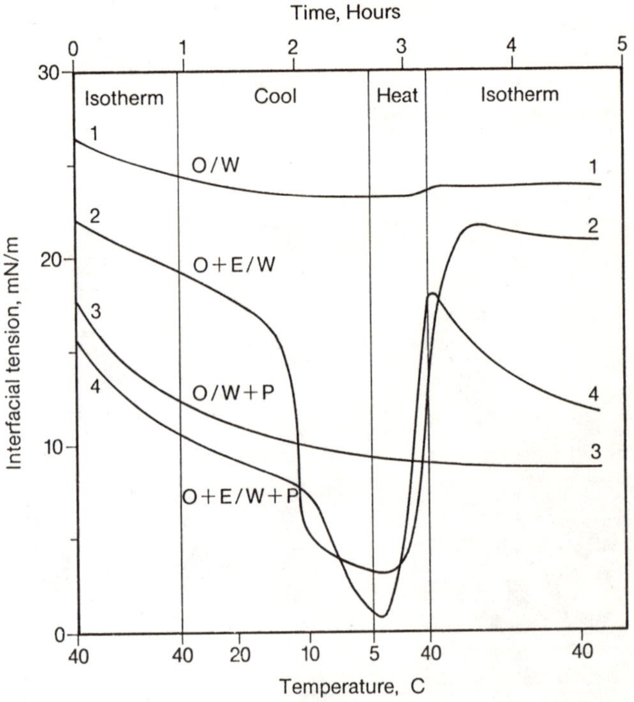


Figure 3. Changes in interfacial tension of oil:water interface with and without emulsifiers or proteins measured as a function of temperature. O = oil phase (sunflower oil), W = distilled water, E = 0.1% emulsifier (GMS or GMO) based on the oil phase, P = 0.01% milk protein based on the water phase.

Figure 3 shows the interfacial tension of various sunflower oil:water systems as a function of temperature. The temperature was first kept at 40°C for one hour, followed by cooling to 5°C at a rate of 0.3°C/min.

When the temperature of the oil:water system had decreased to 5°C, heat was applied via the thermostat. The temperature increased again to 40°C, where it was kept for about 1½ hours. The results show that the pure system of oil:water only drecreases slightly in interfacial tension when the temperature is decreased from 40 to 5°C. With 0.1% GMS added to the oil phase a considerable decrease in interfacial tension is observed when the temperature is below 15°C. At 5°C the interfacial tension is 3-4 mN/m. When re-heated to 40°C, the interfacial tension increases to the initial value of 21 mN/m. With 0.01% milk proteins in the water phase the interfacial tension reaches an equilibrium at 9 mN/m, and is not affected by changes in temperature. A combination of 0.1% GMS in the oil phase and 0.01% protein in the water phase shows that the interfacial tension decreases when the temperature is below 10°C to a value of 1 mN/m at 5°C. After re-heating the interfacial tension increases to a higher value than the initial value of 15 mN/m, which is achieved again after half an hour's equilibration time. The decay in interfacial tension at 40°C is due to the slow re-adsorption of protein.

This experiment shows that GMS becomes much more active as a surfactant when the temperature is lowered below 15°C, and due to this increased surface activity GMS can displace proteins from the oil:water interface. This displacement will take place when the interfacial tension (curve 2) of the emulsifier film intercepts with the interfacial tension (curve 3) of the protein film, which takes place at a temperature of about 12°C in the experiments shown in Figure 3. Recent experiments have shown that GMO performs similarly to GMS at low temperatures.

## DISCUSSION AND CONCLUSION

The relationship between surface tension and temperature in emulsifiers was observed two decades ago by Lutton et al. (10). They explained that this relationship is due to a transition from a liquid-expanded type of mono-layer existing at high temperature (e.g. above 40°C) to a solid-condensed monolayers existing at a lower temperature (e.g. under 20°C). In solid condensed monolayers the molecular packing of the emulsifier molecules is much denser than in liquid expanded monolayers, and these differences result in lower or higher surface tension respectively.

Saturated monoglycerides form liquid expanded monolayers at 40°C, but on cooling to below 20°C a transition to solid hydrocarbon chains takes place, resulting in denser paking of the molecules and a decrease in interfacial tension.

Oil:water systems containing emulsifiers and proteins exhibit lower interfacial tension than systems containing only emulsifiers or protein which are added separately. This means that the interfacial film consists of both emulsifiers and protein units (see Figure 3). On cooling, the interfacial tension decreases in the same way as that of the system containing emulsifiers and no protein. This indicates that the emulsifiers increasingly dominate the interfacial film at low temperature, although some protein may still be present.

If a similar decrease in interfacial tension takes place in an ice cream mix during ageing at 5°C, this decrease will accelerate the desorption of protein from the fat globule surface. This mechanism may explain how monoglycerides promote protein desorption in ice cream mixes and other whippable emulsions.

When measuring the protein desorption from fat globules in ice cream mix aged at 5°C, GMO displayed a stronger effect than GMS, which results in a different texture and melting resistance of the final ice cream (3). The interfacial tension measurements at low temperature shows an increasing adsorption of monoglycerides at the oil-water interface. When the concentration of GMS molecules is increased it results in the formation of solid condensed mono-layers. In case of GMO a formation of a liquid-crystalline mesomorphic phase with cubic structure will take place, and this results in absorption of water into the surface of the fat globules. This mechanism has been demonstrated to take place in other whippable emulsions (11) and may account for the different behaviour of GMS and GMO in ice cream emulsions.

## LITERATURE CITED


1. Anderson, M. and Brooker, B.E. in: Advances in Food Emulsions and Foams, Dickinson, E. and Stainsby, G. (Eds.), Elsevier Applied Science, p. 221, (1988).
2. Buchheim, W., Barfod N.M. and Krog, N., Food Microstructure, 4, 221, (1985).
3. Buchheim, W., Barfod, N.M. and Krog, N., Fett, Wissenschaft Tech., to be published.
4. Goff, H.D., Liboff, M., Jordan W.K. and Kinsella, J.E., Food Microstructure 6, 193, (1987).
5. Darling, D.F. and Birkett, R.J., in: Food Emulsions and Foams, Dickinson, E. (Ed.), Royal Society of Chemistry, London, pp 1-29 (1987).
6. Robson, E.W. and Dalgleish, D.G., J. Food Sci. 52, 1694, (1987).
7. Darling D.F. and Butcher, D.W., J. Dairy Res., 45, 197, (1978).
8. Walstra, P. and Oortwin, H., Neth. Milk Dairy J., 36, 103, (1982).
9. Walstra, P. in: Gums and Stabilizers for the Food Industry. 4th Phillips, G.O. Wedlock, J.J. and Williams, P.A. (Eds.) IRL, Oxford, p. 323, (1988).
10. Lutton, E.S., Stauffer, C.E., Martin J.B. and Fehl, A.S., J. Coll. Interface Sci., 30, 283, (1969).
11. Krog, N., Barfod, N.M. and Sanchez, R.M., J. Dispersion Science and Technology, 10, 483, (1989).

# PHASE SCIENCE OF EMULSIFIERS: SIGNIFICANCE AND RECENT DEVELOPMENTS IN METHODOLOGY

Robert G. Laughlin, Ph.D. ■ Miami Valley Laboratories, The Procter & Gamble Company, Cincinnati, OH 45239

The structure of food products has many qualitatively different dimensions—molecular, conformational, phase and colloidal. Arranging these various kinds of structural information in a hierarchy of relative complexity provides a useful perspective. The phase science of food emulsifiers is directly relevant to the phase structural level, and hence to the thermodynamics of the system. It is also true, but less obvious, that phase behavior influences colloidal structure.

Determining the phase behavior of emulsifiers can be difficult, but both qualitative isothermal swelling methods and the recently developed quantitative analog are readily applicable to food emulsifiers. Exploratory data illustrating the potential of these methods are presented.

The phase science of food emulsifiers is widely recognized as an important dimension of their physical science (1), and extensive phase studies of these compounds using classical methods have been reported.

In the first part of this paper the nature of the information provided by phase information on food emulsifiers, and the relevance of phase science to food technology, will be reviewed. In the second part, exploratory data will be presented which illustrate both the limitations of classical methods and the remarkable potential of swelling methods for producing reliable phase information.

## RELEVANCE OF PHASE INFORMATION TO FOOD EMULSIFIER SCIENCE

To define the physical science of any system one must provide structural, thermodynamic, and kinetic information. Let us consider first "structural information".

The word "structure" is an ambiguous term. Nuclear physicists deal with "atomic structure" and every chemist is familiar with "molecular structure". Organic and polymer chemists deal with "conformational structure," phase chemists with "phase structure", and colloid chemists with "colloidal structure". Biological systems display exceedingly elaborate structures.

In order to perceive the relationships among these different dimensions of "structure", it is useful to arrange them in a heirarchy of relative complexity. Such a heirarchy is shown in Fig. 1.

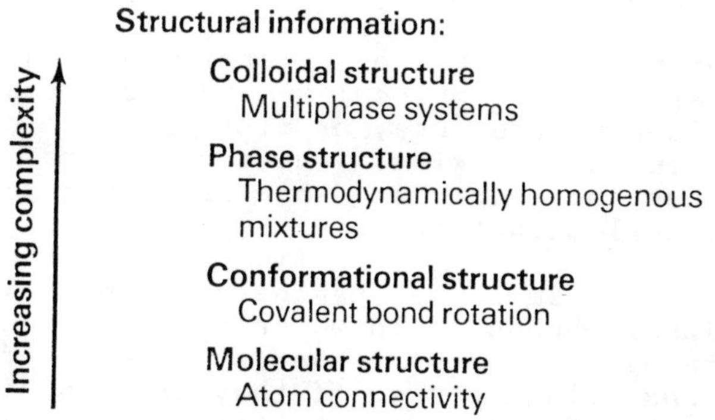


Figure 1. Structural information hierarchy

The simplest kind of chemical structural information is "molecular structure", which is fully defined by stating atom connectivity (which atom is connected to which). Stating molecular structure provides by implication an immense amount of information about a pure substance, but does **not** constitute a full definition of structure - even of a single isolated molecule. This is because practically all molecules display facile isomerism due to rotation about covalent bonds (2). To define the shape of a molecule thus requires conformational structural information. Since more parameters must be stated to define conformational than molecular structure, the former is clearly more complex information than the latter.

These two lower levels of structure suffice to define an isolated molecule, but most chemical investigations involve samples which contain astronomically large numbers of molecules. Such a sample which is thermodynamically homogeneous is termed a "phase" (3).

Phases have characteristic structures. Phase structure may be very simple, as in gases and single crystals. It may be extremely complex, poorly defined, and impossible to describe using a simple drawing, as in liquids (4). Both crystal-like order and liquid-like disorder may exist in the same phase, in which case one has a "liquid crystal" phase (5).

Phase structure describes how aggregates of large numbers of molecules are arranged in space. Within each phase, molecules having a particular molecular and conformational structure, or distribution of structures, exist. Phase structure is thus more complex than is conformational structure.

A sample may be thermodynamically **in**homogeneous, in which case it contains two or more phases. These are coexisting phases if they are in direct contact and in a state of dynamic equilibrium (3). With expenditure of sufficient energy in a proper manner, these various phases may be dispersed within one another in various ways. This level of structure is conveniently termed "colloidal structure". Broadly speaking, colloidal structures fall into one of two classes: sols, which contain discrete particles separated by a continuous matrix, and gels, which contain a network of one phase within a matrix of another (6,7).

In food systems one encounters structural complexity at least to the colloidal level. Simple real systems such as oil/water emulsions in salad dressings, stabilized foams in whipped cream (8), or concentrated dispersions of ice crystals in a mixture of liquid phases in ice cream (9), are reasonably well described using colloidal structure information. In other food systems still higher structural levels must be considered.

Structure is important because it often defines utility. In areas such as pharmaceuticals or molecular biology, molecular structure provides sufficient information to define intrinsic biological activity (10). Indeed modern biology has become synonymous with molecular interpretations of biological phenomena, because a direct correlation between molecular structure and these phenomena often exists.

In the foods area the nutritional value and other properties of foodstuffs **do** correlate with molecular structure. However, for many food properties molecular structure **per se** is inconsequential - except insofar as it anticipates other properties. Two properties of particular interest (with respect to the textural and aesthetic properties of foods (11)) are rheological and optical properties. The correlation between molecular structure and these properties is indirect and empirical. Such empirical relationships between molecular structure and those properties which determine utility apply to many consumer product technologies.

Structural information, in this context, is important mainly as a device for understanding and manipulating utility. Those elements of basic science which underlie and determine structure, and therefore properties, are of particular interest to the food technologist.

## ROLE OF PHASE SCIENCE

The information which underlies phase and colloidal structure is thermodynamics. Thermodynamics is the cornerstone of physical science, because it is independent of rates, path, and mechanism. For the same reasons thermodynamics is by itself incapable of fully defining the physical science of most systems, which do not instantaneously reach equilibrium. As a result, kinetic and other kinds of information are also important.

Further, when a system possesses colloidal structure the rate of change of this structure may be slow. The existence of colloidal structure signifies the existence of an excess energy, commonly termed "surface energy". This kind of instability is diminished by the collapse of colloidal structure, so as to produce coexisting phases having minimal excess surface energy. (The excess surface energy can **never** be zero for coexisting phases, but it may be small in comparison with thermal or mechanical energy fluxes.) It is entirely possible that colloidal structure may never completely collapse.

It follows that there are kinetic aspects both to the attainment of equilibrium at the phase structural level, and to changes in colloidal structure. The two phenomena must be treated separately, because often equilibrium is quickly attained with respect to phase structure while colloidal structure collapses at a much slower rate (12).

For food systems, that branch of thermodynamics which is particularly relevant is phase science. Phase diagrams define, for systems in equilibrium, the range of system variables (temperature, pressure, and composition) over which smooth and continuous changes in phase structure and properties occur. They also define the coordinates of these variables at which discontinuities occur; the loci of these discontinuities are the phase boundaries. Thus phase information is essentially thermodynamic in nature, and is the most elementary form of thermodynamic information.

## CORRELATIONS BETWEEN PHASE EQUILIBRIA AND COLLOIDAL PHENOMENA

Considerable information exists as to just how phase information is relevant to foods in the works of Friberg and Larsson.

Friberg has pioneered investigations of the correlations between phase equilibria and colloid stability. He has documented the fact when small amounts of the lamellar liquid crystal phase coexist with two liquid phases, emulsion stability is dramatically enhanced (13). Further, he has shown that when a foam is created from a mixture of a liquid and a coexisting lamellar phase, it is remarkably stable (14). Thus, it is actually possible to anticipate - by inspection of a phase diagram - the precise compositions which lead either to stability or to instability in foam and emulsion structure. This noteworthy application of phase information had not earlier been recognized.

Larsson has shown that the monoglyceride and polar lipid components of wheat flour lipids play an important role in the stabilization of foam structure during the baking of bread (15). Both gluten and flour lipids contribute to foam stability in this instance. As regards lipids, the formation of a lamellar liquid crystalline phase by compounds such as sugar diglycerides has been shown to contribute to foam stability. Conversely, lipid mixtures which form the hexagonal liquid crystalline phase within bread are comparatively ineffective in stabilizing the foam structure of bread.

The basis for all these phenomena lies in the fact that the lamellar phase - but not the hexagonal or cubic phases - is surface active. This phase spontaneously collects and spreads at interfaces (16). As a result interfacial tensions are lowered, and the drainage of liquid through foam membranes and the Plateau borders (13) at the intersections of these membranes is retarded.

The water-in-oil (L2) microemulsion (13) phases are oily and do **not** form stable emulsions unless the lamellar phase is present. They also diminish foam stability, possibly via Marangoni instability induced by their presence within interfaces (17).

## FLAWS IN EXISTING PHASE INFORMATION

It is evident from the above that phase information on emulsifier-water systems is relevant to food technology. Important early contributions to the phase science of aqueous monoglyceride systems were made by Lutton (18), but it has been discovered that flaws exist in these data. To obtain a correct historical perspective it is worthwhile examining the Lutton diagram of a representative monoglyceride-water system, and the method by which it was determined.

It is evident from the monoolein-water diagram (Fig. 2) that this com-

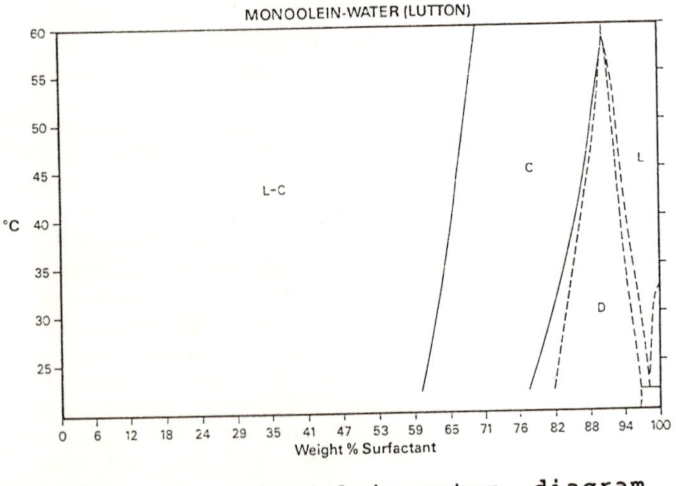


Figure 2. Monoolein-water diagram (Lutton)

pound is poorly soluble in water and that water is soluble in monoolein to about 25%, depending on temperature. Ice and dry monoglyceride crystal phases, plus hexagonal, cubic, and lamellar liquid crystal phases, exist. Crystal hydrates apparently do **not** exist. No Krafft discontinuity (12) is seen over the temperature range of this investigation. The hexagonal phase, designated originally as the "middle" or oil-core phase, has since been found to possess the "inverted" or watercore structure (19).

This diagram was determined using the phase tube method, which is based on the isoplethal phase studies principle (20). (Isoplethal phase studies are performed by varying temperature at specific compositions and determining the temperatures at which discontinuities occur in the number of phases.) For this study numerous samples of known composition were prepared, sealed, mixed by centrifugation through an orifice, and heated and/or cooled while being observed between crossed polars. Phase structures were identified by observation of properties and of textures in the polarizing microscope (21), or by powder x-ray studies.

These studies are laborious and require large samples; further, the results are in some respects uncertain. For example several phase boundaries are dotted, because while they are presumed (on the basis of the Phase Rule) to exist, data which define their precise location do not exist. Indeed, in most surfactant-water phase diagrams (5) those boundaries which define the limits of liquid crystalline phase regions are ill-defined.

Continuing, the manner in which the lamellar liquid crystalline phase decomposes on heating is also uncertain. It is not even clear whether this phase decomposes congruently (to a single phase of the same composition) or incongruently (to two phases of different compositions) (3). Finally, the coexistence relationships of the hexagonal phase with other phases (not shown in Fig. 2) are unclear.

Questions such as these constitute elements of the basic phase behavior of this system which are extremely difficult to define using isoplethal phase study methods, but they are **not** insignificant. These uncertainties signify that the thermodynamics of heterogeneous equilibria in this system have not been fully established - even as to their qualitative aspects.

### Exploratory Monoolein-water Diagram

This same system was recently

reinvestigated during exploratory studies of a wide range of surfactant systems using the new DIT (Diffusive Interfacial Transport) method (22) (Fig. 3). Briefly, the DIT experiment

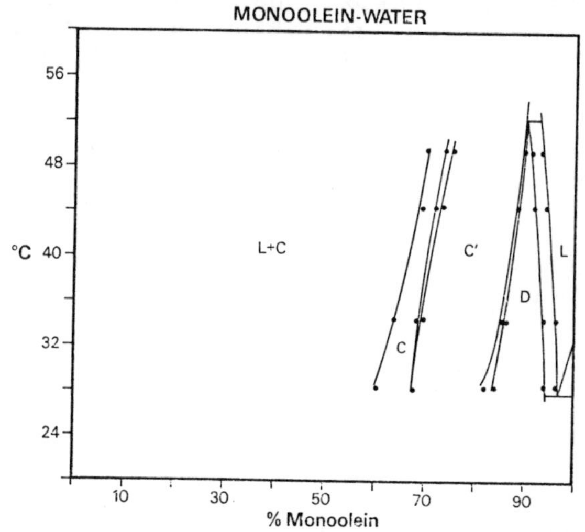


Figure 3. Monoolein-water diagram (preliminary DIT data).

is executed by creating an interface between the compound and water within a long, thin, capillary. After a few hours the swelling of the compound by water (principally via diffusive transport) creates a sequence of bands within the capillary. These bands correspond to the phase sequence at the ambient temperature, and composition discontinuities at the interfaces between the bands correspond to the miscibility gaps between the phases. Qualitative identification of the phase structure within each band, and quantitative determination of the compositions at interfaces, provides phase information.

In this monoolein study the full temperature range was not studied and the raw refractive index data have not been precisely correlated with compositions. However, the qualitative features of the diagram were precisely defined and compositions were estimated from the index-temperature profile of the dry compound, assuming a linear relationship between refractive index and composition. Compositions in the resulting diagram agree with Lutton's data to within ca. 3%, where it is possible to compare the data.

This DIT diagram also shows that the solubility of the compound in water is small, while water is relatively more soluble in the compound. However, two cubic phases were observed, separated by a small miscibility gap.

Only one cubic phase was recognized in the earlier study, which is hardly surprising in view of the narrow miscibility gap between them and the low level of scattering which exists when this phase is dispersed in another isotropic phase (23). (In fairness, it should be noted that the existence of the two cubic phases was recently recognized by Longley and MacIntosh (24) and by Larsson (25), using x-ray data. The convergence of the boundaries of the lamellar phase regions to a sharp cusp suggests that this phase decomposes **incongruently** at a peritectoid discontinuity (12). At the temperature of the cusp the phase disproportionates to a dilute cubic phase and a concentrated liquid phase.

This kind of behavior at the upper temperature limit of liquid crystal phases is not usually observed. Typically, liquid crystals decompose **congruently** to form a liquid phase (5).

Thus, it was possible during the DIT study, using milligrams of sample and about three days' work, to obtain phase information that was either accessible using other methods only with considerable effort (i. e. recognition of all the phases), or not at all (i. e. the compositions of coexisting phases and the nature of the discontinuities).

### The $C_{10}E_4$-H2O System

Another instance in which the use of classical methods led to qualitative errors is the $C_{10}E_4$-H2O system. This system had been exhaustively studied by Lang (26), but we find nevertheless that the published diagram (Fig. 4) is not entirely correct. The phase behavior of this surfactant is unusually complex, due to interaction of the liquid crystal

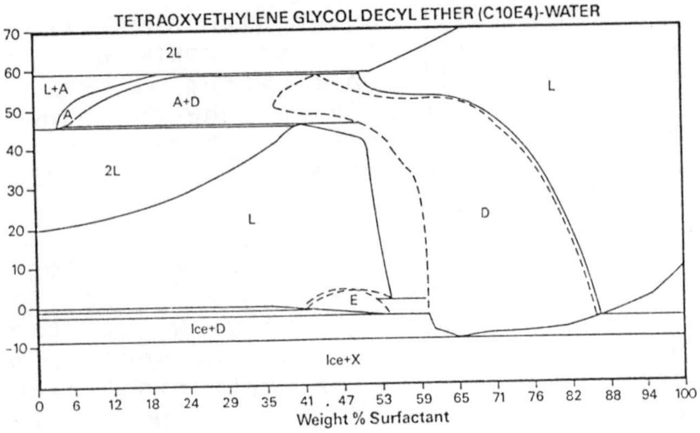


Figure 4. The $C_{10}E_4$-water system (Lang and Morgan)

region with the (cloud point) miscibility gap. This produces a corridor which cuts through the gap and creates a new ($L_3$ or anomalous) phase.

If one scans isotherms in the Lang diagram below 46°C, the phase sequence LLDL is found (using L for liquid and D for the lamellar liquid crystal). Between 46.00 and 46.14°C the sequence LLLDL is observed, and between 46.14 and 58.66°C the sequence LLDL is again found. We find using the DIT method these expected phase sequences - except near the lower part of the corridor. In Fig. 5 a sketch

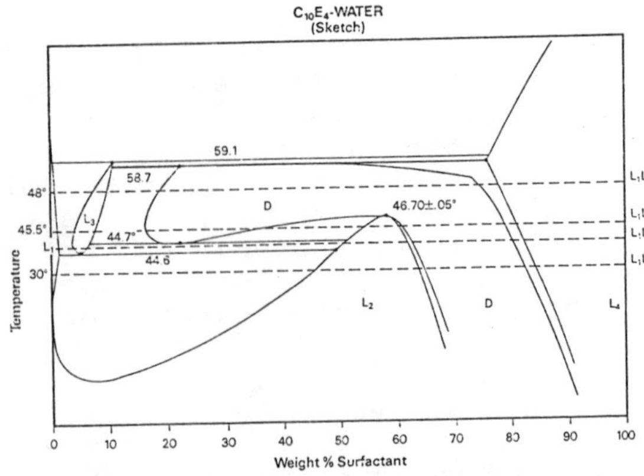


Figure 5. A sketch of the $C_{10}E_4$-water diagram. The temperature scale has been distorted so that the qualitative features suggested by DIT data are visible. The phase sequences along specific isopleths are noted to the right.

of a revised diagram is presented in which the temperature scale is severely distorted. In this way the qualitative features of the phase behavior are visible.

The most significant change is the insertion, above the LLLDL regions, of a **six-phase** LLDLDL region spanning a 2°C range. Nowhere in the published diagram do isotherms exist across which six phases are encountered. This situation results from the intrusion, from below, of the L2 region into the lamellar liquid crystal region. The modification required by the new data is elimination of the cusp at 46% and 46°C in the L2 phase. Instead, the boundary of the L2 phase is smooth and continuous, and the phase terminates at an upper azeotropic point. Such behavior has not previously been observed in surfactant phase studies - and is altogether remarkable. It signifies that **a liquid crystal is formed by heating a micellar solution.** (Actually, this sequence of events does occur, rarely and over narrow composition ranges, when the liquid crystal solubility boundary displays a bulge towards the water axis (5,23).)

## SCOPE AND LIMITATIONS OF SWELLING METHODS

These recent investigations of monoolein and $C_{10}E_4$, as well as other information (27), illustrate the fact that the status of phase diagrams of even the most thoroughly studied systems may be uncertain. They further illustrate the value of using swelling methods during phase studies. These methods will in time become widely used, and it is therefore worthwhile reviewing their scope and limitations. The advantages include:

Efficiency. In each experiment, the phase sequence along an entire isotherm is displayed within a few hours' time. Five minutes are required to quantitatively scan an interface using the DIT apparatus. The calibration of these data presently requires considerably more work, and calibration methods are still being refined.

Thoroughness. The chances of missing a phase are smaller using swelling methods than by any other, because composition is varied in a smooth and continuous manner. In order to insure detection of interfaces it is necessary to observe the system using sensitive interference optics, such as Jamin-Lebedeff (22,28) or Hoffman Modulation Contrast (29). Small refractive index discontinuities may easily be missed using only brightfield and polarized light observation.

Precision and accuracy. When the phases present are either isotropic or precisely aligned anisotropic phases, the accuracy with which compositions can be determined using index data is potentially <0.1% (22). Short tie-lines are as readily determined as long ones.

There are presently several practical limitations to swelling methods.

Reversible melting requirement. The sample must be loaded into a narrow (25 micrometers thick) cell by capillary action. It must therefore melt reversibly (without chemical decomposition) to a low viscosity fluid. Partial studies may often be performed on thermally unstable compounds by loading a low-melting low-water mixture. In such cases that part of the isotherm more concentrated than the mixture is sacrificed, but the balance of the isotherm may be scanned and useful information results. This practice was followed both by us (30) and by Kekicheff and coworkers in their definitive study of the sodium dodecyl sulfate-water system (31,32,33).

Monoglycerides do not typically decompose thermally on melting except in that a 1-monoglyceride may equilibrate with a 2-monoglyceride. The extent to which this isomerization had occurred under the conditions of both of the above phase studies is uncertain.

New, undeveloped method. The DIT method is not fully developed and is not widely available. The analysis of composition is currently based on refractive index data obtained using Jamin-Lebedeff optics, which are not commercially available.

More importantly, it has been found that the analysis of composition using refractive index data is not applicable to all classes of phases (12). It fails in the cases of crystals and many birefringent liquid crystal phases, because these phases do not spontaneously and reproducibly orient within the cell. Investigations using infrared absorption data as an alternative method of analysis are planned.

Complex swelling phenomena. Sometimes swelling does not occur by simple diffusion and clean interfaces do not result. When this occurs particles of one phase may be dispersed within another. This phenomenon was observed during investigations of the dioctadecyldimethylammonium chloride-water system (30), but nondiffusive swelling has not so far been observed with any single-chain surfactant.

## SUMMARY

It has been widely recognized that the phase behavior of food emulsifiers is an important dimension of their physical science. Besides defining the structure (and therefore properties) of the phases which exist, it is relevant to the colloid science of food products. The relationships among the various dimensions of "structure" encountered in this and other areas may be clearly discerned by arranging structural information in a heirarchy of relative complexity.

In developing the DIT phase studies method the original goal was to find an easier and faster way to do phase studies on surfactant systems, but after several years experience our perspective has changed. The application in recent years of the DIT method by us, and of qualitative swelling methods by others, has revealed significant flaws in the phase diagrams of many historically well-studied aqueous surfactant systems (27). This means that we have all along been unaware even of the existence of phases which **must** have been encountered earlier during the processing and use of food emulsifiers. Obviously, the role of these phases in food technology is presently unclear.

Perhaps, then, swelling methods are in fact not a luxury at all. Rather, it may be essential to include them as a matter of course in order to guarantee the production of accurate and reliable phase data.

LITERATURE CITED


1. Larsson, K., Food Emulsions, S. Friberg, Ed., Marcel Dekker, Inc., New York 1976, pp 39-66.
2. Eliel, E. L., Stereochemistry of Carbon Compounds, McGraw-Hill Book Company, Inc., New York, 1962.
3. Glasstone, S., Textbook of Physical Chemistry, D. Van Nostrand Co., Inc., New York, 1946.
4. Temperley, H. N. V., Rowlinson, J. S., and Rushbrooke, G. S., Physics of Simple Liquids, Wiley Interscience Division, John Wiley & Sons, New York, 1968.
5. Ekwall, P., Advances in Liquid Crystals, G. H. Brown Ed., Academic Press, Inc., New York, 1975 ; pp 1-142.
6. Colloid Science, Irreversible Systems, H. R. Kruyt, Ed., Elsevier Publishing Co., New York, 1952.
7. Colloid Science, Reversible Systems, H. R. Kruyt, Ed., Elsevier Publishing Co., New York, 1952.
8. Mulder, H., and Walstra, P., The Milk Fat Globule, Centre for Agricultural Publishing and Documentation, Wageningen, the Netherlands, 1974.
9. Berger, K. G., Food Emulsions, S. Friberg, Ed., Marcel Dekker, Inc., New York, 1976, pp 141-213.
10. Oxender, D. L., and Fox, C. F., Protein Engineering, Alan R. Liss, Inc., New York, 1987.
11. Bourne, M. C., Food Texture and Viscosity: Concepts and Measurements, Academic Press, New York, 1982.
12. Laughlin, R. G., Munyon, R. L., Fu, Y.-C., and Fehl, A. J., J. Phys. Chem., in press.
13. Friberg, S., J. Soc. Cos. Chemists 30, 309-319 (1979).
14. Friberg, S., Advances in Liquid Crystals, G. H. Brown, Ed., Academic Press, Inc., New York, 1978, pp 149-165.
15. Larsson, K., Chemistry and Physics of Baking, Special Publication No. 56, J. M. V. Blanshard, P. J. Frazier, and T. Galliard, Eds., The Royal Society of London, Burlington House, London W1V 0BN, 1986, pp 62-74.
16. Kislalioglu, S., and Friberg, S., Theory Pract. Emulsion Technology, Proc. Symp., 1974, Alec L. Smith, Ed., Academic Press, London, 1976, pp 257-269.
17. Davies, J. T., and Rideal, E. K., Interfacial Phenomena, Academic Press, New York, 1963.
18. Lutton, E. S., J. Am. Oil Chemists Soc. 42, 1068-1070 (1965).
19. Krog, N., and Lauridsen, J. B., Food Emulsions, S. Friberg, Ed., Marcel Dekker, Inc., New York, 1976, pp 67-140.
20. Hill, A. E., J. Am. Chem. Soc. 45 1143 (1923).
21. Rosevear, F. B., J. Am. Oil Chemists Soc. 31, 628-638 (1954).
22. Laughlin, R. G., and Munyon, R. L., J. Phys. Chem. 91 3299-3305 (1987).
23. Laughlin, R. G., J. Coll. Interface Sci. 55, 239-241 (1976).
24. Longley, W., and MacIntosh, J., Nature 303, 612-614 (1983).
25. Larsson, K., Nature 304, 664 (1983).
26. Lang, J. C., and Morgan, R. D., J. Chem. Phys. 73, 5849-5861 (1980).

27. Laughlin, R. G., J. Am. Oil Chem. Soc., submitted.

28. Lebedeff, M. A.-A., Rev. d'Optique 9, 385-413 (1930).

29. Miller, C. A., Ghosh, L., & Benton, W. J., Coll. Surf. 19, 197-223 (1986).

30. Laughlin, R. G., Munyon, R. L., Talmon, Y., Burns, J., and Chestnut, M. H., to be published.

31. Kekicheff, P., and Cabane, B., J. Phys. 48, 1571-1583 (1987).

32. Kekicheff, P., Grabielle-Madelmont, C., and Ollivon, M. J., Coll. Int. Sci. 131, 112-132 (1989).

33. Kekicheff, P., J. Coll. Int. Sci. 131, 133-152 (1989).

# FORMULATION CONSIDERATIONS IN EMULSION PREPARATION AND STABILITY

John L. Cavallo ■ Kraft General Foods, Tarrytown, NY 10591
David L. Chang ■ Clairol, Inc., Stamford, CT 06902
Fouad Z. Saleeb ■ Kraft General Foods, Tarrytown, NY 10591

Emulsification allows compounding immiscible components in a formulation into a single macrophase mixture. Adjustment in the formulation to fine tune the emulsion characteristics is possible, but such an adjustment should not affect product attributes. The required shelf life demands long term emulsion stability such that the functionality and aesthetic characteristics remain unchanged with time. Options available to the formulator to improve stability, possible side effects of actions taken and methods of assessing emulsion stability, from a product development standpoint, are evaluated and presented.

## INTRODUCTION

Emulsions are one of the largest industrial products that have increased tremendously in volume in recent years. Food emulsions are probably on top of that list followed by cosmetics. Emulsion systems are extensively used in many industries such as the pharmaceuticals, agrochemicals, paint, detergents, dry cleaning, and road applications (bitumen emulsion) petroleum, etc.

Emulsions are a class of disperse systems made up of two immiscible liquids, one is the disperse phase and the second is the dispersion medium. The two most common types of emulsions are the oil/water (o/w) and water/oil (w/o) emulsion. However, dispersions of polar oils (e.g., ethylene glycol) in a non-polar hydrocarbon produces and O/O emulsion. In addition, multiple emulsions (w/o/w or o/w/o) are finding many new applications. Apart from a few exceptions, to disperse a liquid into another immscible liquid it is generally necessary to introduce a third component which is capable of adsorbing at the interface between the two liquid phases and help in the emulsification process and stabilize the formed emulsion. Four major classes of stabilizer can be used namely; a) surfactants, b) macromolecules, c) fine particles and d) simple electrolytes. In most cases, combination of two or more emulsifiers is needed to prepare stable emulsions.

Correspondence concerning this paper should be addressed to John L. Cavallo.

In this article the term emulsion is restricted to traditional macroscopic emulsions where the dispersed droplets are generally 0.1μm in diameter and excludes the so called microemulsions (sometimes called swollen micelles). Thermodynamically, emulsions are unstable systems. Ultimately one would expect the emulsions to break i.e., to separate into two liquids since that is the stable lowest free energy. In many instances emulsion are prepared that can be stable, for long periods. This is usually a desirable attribute to improve the shelf life of emulsion based consumer products (e.g., in most food categories). In many instances, the stability of an emulsions must be destroyed to facilitate phase separation (pollutants, creaming, etc.). In other cases, as in the manufacture of butter and cheese, the form of the emulsion is changed from o/w to w/o (emulsion reversal).

It should be mentioned that although considerable progress has been made to form and understand these complex systems, it must be remembered that there remain some quite reproducible but decidedly strange behavior patterns displayed by these systems, in particular, multicomponent systems that are typified by many food products. Two types of instability are commonly associated with an emulsion: Flocculation, a reversible aggregation of the dispersed phase droplets and eventual separation in a gravitational field, i.e., creaming, with the droplets retaining their integrity. The second instability is coalescence, an irreversible fusion of droplets to reduce the interfacial area, hence the interfacial free energy of the system. The emuslion thus separates into discrete bulk phases.

To predict emulsion stability under adverse conditions that are encountered during shipping and storage (such as temperature variations, shaking and vibrations), methods involving accelerated aging at high temperature and high gravitational force have been developed. Judicious adaptation of these methods may offer some indications to the formulator about the system stability, but none is sufficient to guarantee the outcome in real time. Still, these tests are necessary for the formulator in order to establish some system stability ground, since real time storage/shipping evaluation of every formulation attempted is impractical.

In the past many methods have been applied to the determination of emulsion stability. Emulsion systems are exposed to freeze-thaw cycles and their rate of separation determined as a function of both time and temperature. As with centrifugation measurements, freeze-thaw data must be interpreted in terms of actual performance requirements and are best applied to systems whose stability patterns are well established. Classical viscosity measurements also give an indication as to emulsion stability. This technique is best utilized when the measurements are recorded as a function of time. Changes in the system's viscosity indicate droplet coalescence and phase separation. In w/o emulsion systems droplet interaction can be investigated by electrical conductivity measurements. An increase in the conductivity indicates droplet aggregation leading to phase separation. The methods mentioned above have the limitation that they require storage time and multiple measurements in order to get accurate information on the stability of the emulsion.

This article provides a general review of several options that may be taken by the formulator to improve system stability, and several methods of assessing emulsion stability.

## Emulsion Stability Enhancement

### 1. Surfactant Concentration

The role of the emulsifying agent, which is by definition, a surface active ingredient is to accumulate around the dispersed phase droplets forming an interfacial film and lower the interfacial tension. In addition, the interfacial film will prevent or retard particle colaescence and flocculation. Since an emulsion is inherently unstable, the final kinetic stability of the emulsion will be largely determined by this surfactant film. Therefore, a sufficient amount of surfactant must be present in the formulation to cover all the newly created interface resulting from dispersion. Not only is the quantity of surfactant important, but how fast it adsorbs, packs, and orients at the interface must be considered. During emulsification, mechanical energy is supplied to breakdown the dispersed phase and transport surfactant molecules to the newly created surfaces.This essential for both emulsion formation and subsequent emulsion stability. Three major surfactant types are available for emulsifications: a) ionic surfactants (anionic and cationic), b) non ionic surfactants and, c) Zwitterionic (e.g. lecithins). In many instances, emulsion instabilities are sensitive to the concentration of the emulsifier, particularly that of nonionic surfactants (1). There exists a critical emulsifier concentration above which the breakdown of the emulsion (flocculation or coalescence) would occur at an accelerated rate.

The minimum amount of surfactant needed for a particular system may be calculated by considering the total interfacial area formed and the surface area occupied per surfactant molecule (2). For a given volume V (ml) to be disperse into spherical droplets of radius $r$ (Å), the expression for V may take the form

$$V = a\left(\frac{4}{3}\right)\pi r^3 \quad (1)$$

where a is the total number of droplets formed due to dispersion.

The total surface area A of the droplets is

$$A = a\,4\pi r^2 \quad (2)$$

or

$$A = n\sigma \quad (3)$$

where n is the number of surfactant molecules and $\sigma$ is the area per surfactant molecule. Combining Eqs. 1, 2 and 3 and solving for n:

$$n = 3V/r\sigma \quad (4)$$

Equation 4 provides the minimum amount of surfactant needed for complete interfacial coverage. A correction factor in the amount of the CMC of the surfactant is added. This is to account for the non-adsorbing surfactants and other oligo-aggregates that may form in solution. The value of $\sigma$ can be obtained from the surface tension vs. log concentration plot of the surfactant solution. The upper limit for $r$ is of the order of 2,500 Å for a kinetically stable emulsion droplet (3).

Commercial emulsions generally contain a larger excess of surfactants than is needed to form a tightly packed monolayer for complete interfacial coverage. Very often, a reduction in the amount used would compromise system stability. This type of instability may be caused by sub-optimum HLB used for the system (discussed in the next section), and/or a weakening of the interfacial film that prevented particle interaction. To strengthen the interfacial film, multilayer barrier in the form of liquid crystals may be necessary for droplet stabilization (4). Thus, a higher surfactant concentration is required for the system. However, it should be noted that the presence of free surfactant in the emulsion may cause the system to flocculate. This destabilizing effect is said to be induced by exclusion of surfactant micelle between approaching oil droplets (1).

## 2. Selection of Surfactants

A suitable emulsifier system is needed in an emulsion to keep the phases dispersed. The selection of different surfactants in the preparation of different emulsions is often still made on an empirical basis. Very often, the semi-emprical concept of hydrophilic-lipophilic balance (HLB) is used in the selection of the emulsifier. Table 1 lists the HLB requirements of some common oils in o/w or w/o emulsion systems (generally a high HLB for o/w emulsion and a low HLB for w/o emulsions).

The HLB number of a given ionic surfactant may be altered by, for instance, changing the salt concentration of the emulsion system, provided that such a change does not render the final product unacceptable. For example, with increasing NaCl concentration in the system, the solubility of sodium lauryl sulfate (HLB=40) would decrease (salting out), and the surfactant becomes more lipophilic. Figure 1 is a schematic representation of the phase behavior showing the effect of added salt.

It has been reported that certain electrolytes are capable of greatly enhancing the high temperature stabilizing ability of the emulsifiers (5). Table 2 lists some examples of the effects of electrolytes on the high temperature (125°F) stability of water-in-mineral oil high internal phase emulsions. The emulsions contain 91% aqueous phase by volume, and 20% emulsifier concentration based on the volume of the oil phase. The HLB number of ethylene oxide-containing nonionic surfactants is often related to their phase behavior in water (6,7). Table 3 lists some selected nonionic surfactants along with their HLB values.

Shinoda and co-workers (8,9) found that many oil in water emulsions stabilized with non-ionic surfactants undergo a process of inversion to w/o emulsion (the phase inversion temperature, PIT). It is a very useful tool to help in selecting the surfactants with the right HLB for a given emulsion stability. Also, an advantage of this phase inversion temperature is that when such an emulsion is cooled rapidly, the reformation of the o/w system will have smaller droplet size. In practice, emulsifiers are chosen such that the PIT values are well above the "typical" storage temperature of the emulsion system.

Beerbower and Hill (10) introduced a more quantitative index for selecting emulsifiers. It is based on the cohesive energy ratio concept of the mixtures of oil with the hydrophobic portion of the surfactant as compared with the water with the hydrophilic portion. The major draw back of the method is the lack of enough thermodynamic data for surfactant interactions in water and various oils.

3. Viscosity and Particle Size

The physical stability of an emulsion is given by Stoke's law (7):

$$V = \frac{d^2(\Delta P)g}{18n} \quad (5)$$

where

- V = rate of rise (for o/w emulsions) or fall (for w/o emulsions) of the droplets
- d = droplet diameter
- $\Delta P$ = density difference between external and internal phases
- g = gravitational constant
- n = viscosity of external phase

And the average mass creaming rate $\bar{V}$ may be predicted by:

$$\bar{V} = \sum \frac{8\pi}{27nL} g n_i r_i^5 \Delta P \quad (6)$$

where

- L = volume of internal phase
- $n_i$ = number of droplets of radius $r_i$

According to Eqs. 5 and 6, emulsion stability may be improved by:

(1) Reducing particle size and narrowing particle size distribution.
The particle size influences both rate expressions significantly. The particle size may be manipulated by choice of emulsifier (and/or blend), concentration of emulsifier, and order of addition. In practice, smaller particle size and more uniform particle size distributions are usually obtained when the emulsion goes through a phase inversion stage during preparation, i.e., start from a w/o coarse emulsion to produce a final o/w emulsion.

2) Minimize density difference of the two phases.
This is one factor that the formulator has less control of. If the densities can be made the same, then separation would not occur. In practice this balancing method is only applicable in few cases.

3) Increasing the viscosity of external phase.
The rate of separation may be retarded by using additional viscosity imparting agents. The use of thickeners is widely used in practical emulsions. These are usually high molecular weight substances of natural or synthetic type e.g., xanthan, guar and carrageenan gums, as well as various chemically modified cellulose gums.

All these materials when dissolved in the continuous phase increase the viscosity of the medium and hence reduce the creaming or sedimentation rate. In most instances these materials give non-Newtonian solutions which are viscoelastic at fairly low concentrations in water. It is therefore, useful to define the shear rate at which that viscosity is measured for predicting the sedimentation rate. Finely divided solid particles have long been known to stabilize o/w or w/o emulsions (11,12). The mechanism of stabilization of emulsion drops is not fully understood at present, but the presence of finely divided solid particles at the liquid/liquid interface, presumably, plays an important role in preventing the thinning of the liquid films between the droplets. The contact angle at the three phase (solid/oil/water) boundary is critical in controlling the stabilization by a given solid particle and to determin whether an o/w or w/o emulsion is produced (12). Generally speaking rough asymmetric particles such as bentonite clays are more effective than smooth spherical particles.

In summary and particularly for food emulsions, there are numerous natural and artificial emulsifiers and stabilizers of varying molecular weights and chemical properties to choose from. However, many federal and local regulations limit the selections and levels allowed but more important the quality of the food product is even more critical for the consumer acceptance. Many proteins and proteinacous material are utilized in food emulsions, but there is not enough published surface chemical data available for commercially produced proteins which are generally not pure materials and vary from one lot to the other. The same is true for various gums available in the market. The adsorption mechanism of many of these macromolecules at the liquid-liquid interface is less defined than that known for solid/liquid interfaces. At the liquid-liquid interface, segments of a given polymer chain will partition themselves between the two liquid phases in contrast with the solid/liquid interface where the molecule adsorbs with a small fraction of segments in close contact with the solid surface (trains) leaving loops and long dangling tails in the continuous phase (13). These arguments are in support of our introductory remarks that preparation and production of immense volumes of many food emulsion still rely on art and trade secrets in addition to known and understood technical/basic information available to the (public) scientific community.

Methods of Assessing Emulsion Stability

Over the years, many techniques of varying complexities were developed to characterize emulsion and to assess their long term stability. Among these techniques are:

i) interfacial tension measurements including DuNouy's ring, Wilhelmy plate, pendent drop, sessile drop, spinning drop and the drop volume or drop weight method.
ii) the interfacial pressure method (spread monolayers)
iii) the interfacial elasticity
iv) interfacial viscosity
v) Flocculation rates as determined by light scattering techniques, drop counting methods or centrifugal measurements
vi) coalescence rate as measured by following droplet number and/or size with time
vii) rheology of emulsion including steady state measurements (creep and stress relaxation) and dynamic (oscillatory) data. Of more significance to follow is rheological changes of emulsions on storage.

The information obtained from the above mentioned liquid/liquid interface and properties of the bulk emulsion was used by many investigators to successfully predict the long-term physical stability of various emulsions. It is beyond this article to go in details in discussing these techniques. However, we would like to discuss two techniques that were used successfully in characterizing and predicting stability of a number of consumer products, namely pulsed NMR and differential scanning calorimetry. The techniques are simple but informative as they measure properties associated with structure/properties of the interfacial region need to stabilize an emulsion.

1. Pulsed NMR

Trumbetas et al. (14) demonstrated that proton NMR can be used for measuring the extent of oil solidification during cooling of an o/w emulsion. The extent of solidification is directly related to emulsion stability. They established that unstable emulsions would show little supercooling, whereas those that were relatively stable to creaming and phase separation were resistent to oil solidification.

The authors (14) generally believed that dispersions of fat into small droplets bring them into more intimate contact with the aqueous phase, and this would have an inhibiting effect on its rate of solidification during cooling. Since the greater the degree of dispersion the slower the rate of phase separation, pulsed proton NMR is a useful technique to investigate emulsion systems, because it can differentiate between the oil and aqueous phase, resulting from large differences in the relaxation time between oil and water protons due to saturation of the aqueous phase. To use this technique, the oil and aqueous phases of the emulsion is studied individually. Each phase is slowly cooled from room temperature and the NMR signal recorded as a function of time and temperature. The NMR signal obtained for the aqueous phase is very small indicating rf saturation. The lipid phase containing the lipophilic emulsifier behaves quite differently. Upon cooling the NMR signal response increases until solidification begins. The signal then decreases as the sample solidifies. If the process of emulsification had no effect on the rate and extent of fat solidification, the signal obtained for an emulsion system would be approximately additive (the sum of the aqueous and oil phases). However, the signal obtained is always larger than predicted, indicating that emulsification has an inhibiting effect on fat solidification.

To illustrate this effect, an emulsion containing 48% hydrogenated oil, 1% acetylated monoglyceride (49% total fat phase), 1% Tween 20 and water (51% aqueous phase) was investigated (Figure 2). NMR measurements on the aqueous phase show that it gives a very small signal as expected during the entire cooling cycle. As the fat phase is cooled, the signal response increases and then decreases on solidification. This data allows the "percent interaction" to be determined as follows:

| | | |
|---|---|---|
| Fat phase signal = 71.5 x 0.49 | = | 35.0 |
| Aqueous phase signal = 7.51 x 0.51 | = | 3.8 |
| Expected emulsion signal | = | 38.8 |
| Observed emulsion signal | = | 41.5 |
| % interaction = $\frac{41.5 \times 100}{38.8} - 100$ | = | 7.0 |

A correlation was made between the extent of interaction and emulsion stability. If the amount of interaction is large, the resistance to phase separation is increased. In general, if the calculated percent interaction is greater than 30%, the emulsion is considered to be relatively stable. This technique can also be adapted to surfactant blend optimization. At the optimum HLB value for a particular oil, interactions as determined by pulse NMR cooling curve measurements is at a maximum demonstrating higher system stability.

2. Differential Scanning Calorimetry

The moisture content of emulsions is frequently described as consisting of two portions: bound and free. The meaning of bound water is taken as that portion of the total moisture which has measurable properties that differ from those of bulk water or simple solutions (15). These differences may be attributed to interaction forces between the water molecules and the various "solid" components of the emulsion.

Bound water in emulsions can be determined by DSC as the amount of nonfreezable water within a sample after it is cooled to at least -70°C with liquid nitrogen. When the temperature of the cooled sample is increased at a constant rate in the calorimeter, the fusion of ice is detected as an endothermic peak with an area proportional to the amount of ice present. Nonfreezable water is simply the difference between the total water content and the amount of water detected by the fusion endotherm.

In w/o emulsions systems where the droplet diameters may vary, the freezing characteristics of the dispersed water phase may indicate whether the emulsion will have any long term stabilty. When an w/o emulsion is cooled in a DSC, the crystallization of water in most of the droplets results in an evolution of the latent heat when the temperature reaches -40°C (the homogeneous nucleation temperature), yielding an exothermic peak in the thermogram. For droplet diameters greater than about 5μm, a second exothermic peak is observed in the thermogram at a higher temperature. In emulsions with droplet diameters smaller than 5 μm, the heterogeneous nucleation crystallization is not observed (16). Figure 3 curve A shows the two heterogeneous nucleation exothermic peaks for a w/o emulsion containing n-heptane/Span 65 (sorbitan tristearate)/water with droplet diameters greater than 5μm. Curve B shows the single homogeneous nucleation peak when the droplet diameters are 5μm or less. Curve C shows the heating curve for emulsion B, the endothermic peak at 0°C is the melting of ice in the continuous phase. These measurements are essentially related to the polydispersity of the emulsion systems, therefore the use of DSC can give an indication on the extent of droplet coalescence (exhibiting more bulk water characteristics on the thermogram) hence the stability of the emulsion system.

It has also been shown that the kinetics of globule coalescence in w/o and o/w emulsions stabilized by nonionic emulsifiers are influenced by HLB values of the emulsifiers used. Shinoda (17) has demonstrated that the phase inversion temperature (PIT) of o/w emulsions stabilized by nonionic emulsifiers is influenced by emulsifier HLB. Therefore, it appears that there exists a relationship between the kinetics of droplet coalescence, emulsion stability and PIT which may be useful in predicting the stability of emulsion systems.

This technique relies on the use of differential thermal analysis (DTA) to evaluate the PIT of freshly prepared emulsions composed of a combination of nonionic emulsifiers at various HLB values (18). Once the optimal HLB value is determined for a particular oil, the rate of droplet coalescence is evaluated as a function of temperature. It is readily demonstrated that for each emulsifier blend the maximum stability (i.e. the minimum rate of droplet coalescence) coincides with the highest PIT, so that more thermal energy is required to break down the most stable emulsions. The highest PIT values are associated with those emulsions which adsorb surfactant molecules at the oil/water interface in the form of a mixed molecular association complex (19). The results obtained indicate that the phase inversion of o/w emulsions stabilized with nonionic surfactant is an endothermic process. Emulsions formulated at the optimal HLB will have higher PIT values and will be inherently more stable. This method of testing the stability of emulsions systems is a rapid evaluation that requires very little sample (ca. 15 mg) and can be a useful tool to the formulator.

## Conclusion

Many physical chemical methods have been used to evaluate and improve the stability of emulsion systems. Improvements have relied on a more fundamental understanding of raw materials, optimization of processing conditions and a deeper understanding of surfactant behavior in solution which include; the formation of mixed liquid crystal films, phase inversion and HLB optimization. These improvements have led to the development of accelerated stability test methods and new concepts for emulsion systems which have been brought to the market place in the form of products.

Methods that accelerate the determination of emulsion stability are a beneficial tool to the formulator of both cosmetic and food based emulsion products. The advantage of the techniques described include: they reduce the amount of work needed for an accurate evaluation to be made and they often give insight into the fundamental factors governing emulsion stability.

## REFERENCES


1. Aronson, M.P., Langmuir, 5, 494 (1989)
2. Rosano, H.L., J. Colloid Interface Sci., 44, 242 (1973)
3. Bhargava, H.N., Narurkar, A., and Lieb, L.M., Pharm. Technol. 11, 36 (1987)
4. Pilpel, N., and Rabbani, M.E., J. Colloid Interface Sci., 122, 266 (1988)
5. Aronson, M.P., and Petko, M.F., U.S. Patent #4, 606, 913 (1986)
6. Krog, N.J., Riisom, T.H., and Larsson, K., in "Encyclopedia of Emulsion Technology," Vol. 2, Becher, P., ed., Marcel Dekker, New York (1985)
7. Mitchell, D.J., Tiddy, G.J.T., Waring, L., Bostock, T., and McDonald, M.P., J. Chem. Soc., Faraday Trans. I 79, 975 (1983)
8. Shinoda, K., J. Colloid Interface Sci., 25, 396 (1967).
9. Shinoda, K., and Saito, H., J. Colloid Interface Sci., 30, 258, (1969).
10. Beerbower, A., and Hill, M.W., Amer. Cosmet. Perfum. 87, 6, 85, (1972).
11. Pickering, S.V., J. Chem. Soc., 112, (1934).
12. Schulman, J.H., and Leja, L., Trans. Faraday Soc. 50, 598 (1954).
13. Tadros, Th.F., "Effect of Polymers on Dispersion Properties", Academic Press, London/New York, (1982).
14. Trumbetas, J., Fioriti, J.A., and Sims, R.J., JAOCS 53,722 (1976)
15. Kuntz, I.D., and Kaugmann, W., Adv. Protein Chem., 28,239 (1974)
16. Bruggeller, P.J., Dispersion Sci. and Technology, 3 (4) 395, (1982)
17. Shinoda, K., J. Colloid Interface Sci. 30, 2, (1969)

18. Matsumoto, S., and Sherman P., J. Colloid Interface Sci. 33, 294, (1970)

19. Parkinson, C., Sherman, P., J. Colloid Interface Sci. 41, 328, (1972)

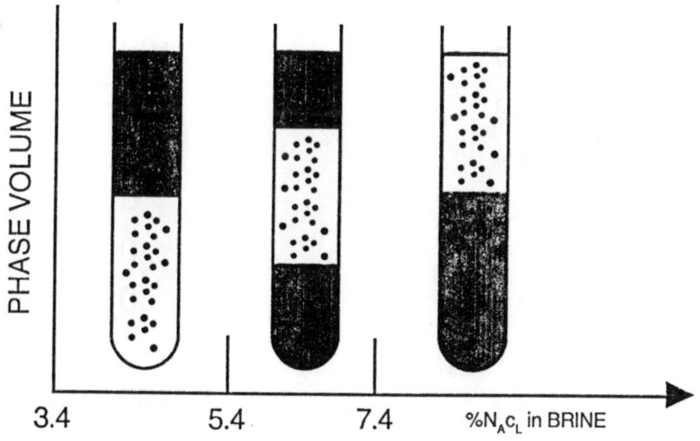


Figure 1. Phase transition in the quaternary mixture water (brine, 46.8%), oil (toluene, 47.2%), SDS (2%), and butanol (4%), as a function of salt concentration in brine.

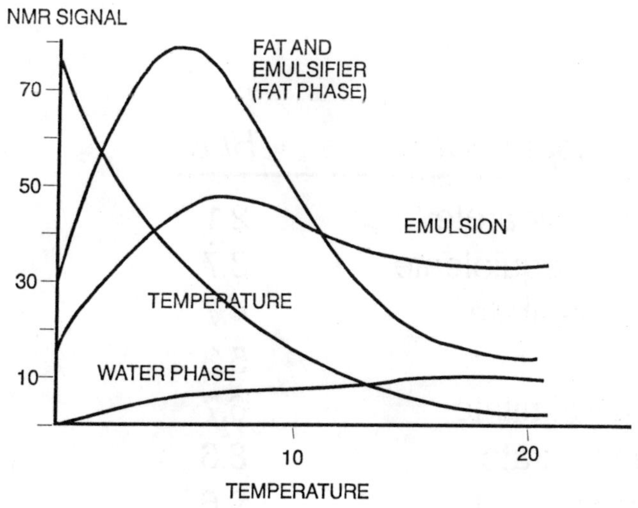


Figure 2. Cooling curves on an emulsion and on its individual phases.

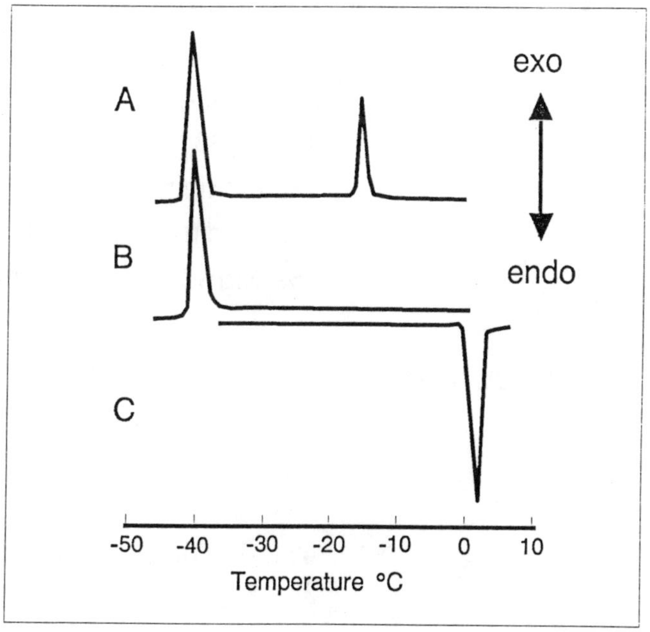


Figure 3. Thermograms for emulsified deionized distilled water; A: Cooling of an emulsion with droplet diameters > 5 um; B: Cooling of an emulsion with droplet diameters < 5 um; C: Warming curve of emulsion B.
(Bruggeller, P., J. Dispersion Sci. Technol. 3,395 (1982)).

Table 1. Required HLB values of some common oils in w/o and o/w emulsions.
(Griffin, W.C., J. Soc. Cosmet. Chem., 5, 249, 1954).

| OIL TYPE | W/O EMULSION | O/W EMULSION |
|---|---|---|
| Mineral Oil, aromatic | 4 | 12 |
| Mineral Oil, paraffinic | 4 | 10 |
| Vegetable Oil | – | 7-12 |
| Vitamin Oil | – | 5-10 |
| Castor Oil | – | 14 |

Table 2. Effect of electrolytes on the high temperature (125 °F) stability of water-in-mineral oil high internal phase emulsion systems. Atmos 300: mono and di glycerides of fat forming fatty acids. Drewpole 10–4–0: Decaglycerol tetraoleate. (Aronson, M.P., and Petko, M.F., US Patent #4,606,913, 1986).

| Emulsifier | Electrolyte | Time of Storage (Days) | % Separations of Water |
|---|---|---|---|
| ATMOS 300 | Distilled Water | 49 | 100 |
| | 1% $MgCl_2$ | 49 | 0 |
| | 1% $K_2SO_4$ | 49 | 0 |
| | 6% Sodium Lactate | 49 | 0 |
| | 1% Sodium Polyacrylate | 49 | 0 |
| | 1% $MgSO_4$ | 115 | 0 |
| DREWPOL 10-4-0 | Distilled Water | 49 | 100 |
| | 6% Sodium Lactate | 49 | 10 |
| | 4% Sodium Polyacrylate | 49 | 0 |
| | 4% $AlCl_3$ | 49 | 0 |
| | 4% Sodium Polyacrylate | 200 | 0 |
| | 4% $MgSO_4$ | 200 | 0 |

Table 3. HLB values of some nonionic surfactants.

| Product Name | CTFA Adopted Name | HLB |
|---|---|---|
| SPAN 65 | Sorbitan Tristearate | 2.1 |
| ARLACEL 83 | Sorbitan Sesquioleate | 3.7 |
| SPAN 60 | Sorbitan Stearate | 4.7 |
| BRIJ 52 | Ceteth — 2 | 5.3 |
| SPAN 40 | Sorbitan Palmitate | 6.7 |
| SPAN 20 | Sorbitan Laurate | 8.6 |
| TWEEN 61 | Polysorbate — 61 | 9.6 |
| TWEEN 81 | Polysorbate — 81 | 10.0 |
| TWEEN 85 | Polysorbate — 85 | 11.0 |
| BRIJ 76 | Steareth — 10 | 12.4 |
| TWEEN 60 | Polysorbate — 60 | 14.9 |
| TWEEN 20 | Polysorbate — 20 | 16.7 |

# THIN FILM STABILITY AND INTERFACIAL RHEOLOGY OF EMULSION SYSTEMS

Lloyd A. Lobo and Darsh T. Wasan ■ Department of Chemical Engineering, Illinois Institute of Technology, Chicago, IL 60616

The stability of food grade emulsions is discussed along with the interfacial rheological properties and dynamic thin film behavior. The systems include soybean oil-in-water emulsions stabilized by a protein and a nonionic emulsifier. The protein is found to cause a large increase in the interfacial viscosity which inhibits film drainage and hence stabilizes the emulsions. On the other hand, emulsions containing the nonionic emulsifier, which does not affect the interfacial viscosity appreciably, are also found to be stable. Draining thin films formed from such a nonionic micellar solution exhibit sharp thickness transitions which is indicative of the stratification phenomenon. These observations corroborate our newly proposed mechanism of emulsion stability, due to the presence of long range structural forces in thin films over the distance of the order of 1000 A and the formation of colloid crystal-like structures within the film.

## INTRODUCTION

Emulsions are an inherent part of many processed foods such as salad dressings, cheese, margarine, processed meats, etc. Most often stable emulsions are desired, to provide quality and maintain a long shelf life for the product. The stability that we refer to here is in the kinetic sense, since emulsions are thermodynamically unstable, because the free energy of the emulsion always decreases with coalescence (1).

The breakage of emulsions involves two processes - 1) flocculation of the drops and 2) coalescence. Once flocculation occurs and the two drops approach each other, a thin film of continuous phase is trapped between the drops. The film drains to a critical thickness and then ruptures. The rate of drainage of the thin films is critical to the stability of the emulsion.

*Illinois Institute of Technology, Chicago, Illinois.

At low surfactant concentrations (cmc or below) the hydrodynamics of film drainage are governed by the interfacial rheological properties such as interfacial viscosity and tension gradients or elasticity. At very high concentrations the surfactant forms mesomorphic phases called liquid crystals, whose viscosity is about 100 times that of water, thereby reducing drainage rates (2). At concentrations substantially higher than the cmc, but lower than those at which liquid crystalline phases are formed, the micelles form an ordered structure within the thin film which inhibits drainage (3,4).

Liquid crystals were observed in emulsions by Friberg et al. (5) and they later showed that liquid crystalline states may form at the surface of oil drops in o/w emulsions and provide stability against coalescence (5). Liquid crystalline phases are formed by many naturally occurring lipids or commercial emulsifiers, but the concentration of emulsifiers in typical food applications is often too low for liquid crystals to form. Furthermore, most food emulsifiers are too lipophilic to form liquid crystal structures of the lamellar type with oil and water at the interface (6). There is little experimental evidence to suggest that liquid crystalline phases actually exist in food emulsions.

In this paper, we studied the stability of soy bean oil-in-water emulsions, which is of interest in food applications. A model protein as well as a food grade emulsifier were used to stabilize these emulsions. We found that the drainage of the thin films, and thereby the stability of these food emulsions, is caused predominantly by the increase in interfacial viscosity and the "ordered" structures observed in such microheterogeneous systems.

## EXPERIMENTAL

### Materials and Methods

The soy bean oil used was a commercial brand manufactured by Beatrice/Hunt Wesson, Inc. In order to eliminate inherent surface active substances, the oil was passed through a bed of Florosil (200 -300 mesh) adsorbent, according to a procedure suggested by Gaonkar (7). This procedure resulted in an increase in interfacial tension, of the oil with distilled deionized water from 19 dyne/cm to 29.9 dyne/cm.

Bovine serum albumin (BSA), with 98-99% purity, was used as the model protein, and was obtained from Sigma chemicals. The food grade emulsifier we used, is a commercial brand sold by Durfax. It is an ethoxylated monoglyceride with an average molecular weight of 1360. The emulsifier was provided by Kraft Inc.

Monobasic and dibasic potassium salts of phosphoric acid were obtained from Sigma chemicals. These were used to prepare a buffer solution. The buffer also contained 0.3 gm/lit of sodium azide (Sigma Chemicals) as a toxin to prevent biodegradation of the protein with time. Solutions of protein and emulsifier were made in a 0.1 M phosphate buffer (pH = 7).

Interfacial Viscosity Measurements: Interfacial viscosity measurements were made using the deep channel viscous traction interfacial viscometer (8-11). The detailed description of the instrument is given in references 10 and 11. A small tracer particle (teflon) is placed at the oil/water interface and the base of the deep channel is rotated. The velocity of the particle, which moves along with the interface, is recorded as a function of time. For all the experiments the depths of the oil and water phases as well as the velocity of the dish were kept constant.

Interfacial Tension Measurements: Interfacial tension for the oil water system was measured using a Wilhelmy plate made of platinum and a Kruss tensiometer.

Preparation of Emulsions: The emulsions, containing 30% v/v of soy bean oil, were prepared using a microfluidizer M110 (Microfluidizer Corp., Newton, MA). Initially, a coarse emulsion was prepared by adding the soy bean oil drop-wise to a continuously stirred aqueous solution containing the stabilizer (protein or emulsifier). The coarse emulsion was passed through the microfluidizer to obtain the fine dispersion. The microfluidizer works on the principle of impinging two high velocity jets of liquid in an interaction chamber. In this way the energy used to create the dispersion is fairly constant for every emulsion prepared.

Emulsion Stability Tests: The emulsions that we studied were stable as there was no measurable separation of the oil and aqueous phases with time. In order to determine emulsion stability, we used the photomicrographic technique employed by Menon and Wasan (12). The emulsion stability is related to the change in the number of drops, per unit volume of the dispersed phase, as a function of time. A sample of the emulsion was placed in a Howard Cell and photographed through a Nikon microscope. The microscope is equipped with phase interference contrast and polarizing light. The first photograph was taken as soon as the emulsion was prepared and subsequent photographs were taken after suitable time intervals. The slides of these photographs were analyzed using the Zeiss MOP image analyzer for the purpose of determining the size of the drops. The number of drops per unit volume of the dispersed phase n is given by

$$n = \frac{\sum_i i}{\sum_i \frac{4}{3}\pi r_i^3}$$

where $r_i$ is the radius of each drop

Thin Film Studies: We used a reflected light differential interference microscope to study thin emulsion films stabilized by protein as well as emulsifier. The schematic of the experimental setup is shown in Fig. 1. The films were formed by floating oil drops at an oil-water interface. The water, containing a surface active agent, was placed in the inner cylinder. The cell was then completely filled with soy-bean oil. The objective (12.5 magnification) was then immersed in the oil. Oil drops, of the order of 0.5 mm, were formed using a microsyringe with a flat tipped needle. As soon as the oil drops were formed the objective was focussed onto the emulsion film to take photomicrographs. By using this technique, the films that we studied were curved and had a high capillary pressure. These films are relevant to the real emulsion systems.

Light Scattering Measurements: In order to determine the size of micellar aggregates of the emulsifier system dynamic light scattering measurements were made using the Malvern Automeasure 4700 system. This system measures particle size by laser light scattering and photon correlation spectroscopy (PCS). It has a 128 channel correlator with a smallest fundamental sample time of 20 nanoseconds and can measure particle sizes from 2 to 5000 nm. The correlator uses variable time expansion which is especially suited to measure samples with large size distribution widths.

## RESULTS AND DISCUSSION

### Protein Films

The interfacial tension isotherm for the protein system is shown in Fig. 2. These measurements showed that the interface is completely saturated at a bulk concentration of $10^{-5}$ wt% of BSA. All the emulsion stability tests were carried out with bulk protein concentrations which were 10,000 times greater than this value.

We also observed that the time taken to reach the equilibrium value of the interfacial tension was about 1 - 4 hours for the concentrations used in the emulsion stability tests. Thus, the protein systems exhibited aging, and the interfacial tension changed with time, even at high protein concentrations. This occurs because of the slow adsorption/diffusion kinetics at the oil/water interface.

The results from the interfacial viscosity measurements are shown in Fig. 3. We see that for all the three concentrations of protein, the velocity ratio (dish/particle) increases, with time, to a very high value which is beyond the measurement range of the instrument. Physically, this means that the interface becomes rigid (highly viscous) and the particle velocity becomes zero. The change in the interfacial viscosity with time is also due to the slow adsorption/diffusion kinetics of the protein at the interface.

In real dispersed systems, interfaces are continuously being stretched or generated. In the case of macromolecular stabilizers, such as protein, the interfacial properties, such as interfacial viscosity and interfacial tension, under dynamic conditions, lie in between the equilibrium value and the value for the pure oil/water system. This deviation from the equilibrium is directly related to the aging time of the interface. As we see from interfacial viscosity measurements in Fig. 3 as well as from our observations with interfacial tension measurements the aging time increases with a decrease in the protein concentration.

Fig. 4 shows the emulsion stability tests for the protein system, with the number of drops plotted as a function of the emulsion life. We see that for all the three protein concentrations studied, the number of drops decreases with time, thus indicating that coalescence occurs with the passage of time. The coalescence is much more rapid when the protein concentration is lower. Wasan and co-workers (13,14) have developed an interfacial rheological model to predict the drainage of thin films. This model shows that interfacial viscosity inhibits film drainage and that an increase in interfacial viscosity results in an increase of the drainage time of the films. In the protein systems, the increase in the interfacial viscosity, combined with the aging of the interface, controls the thin film drainage as well as the emulsion stability. We also see that the initial number of drops are smaller with the decrease in the protein concentration. The initial number of drops generated is a function of the interfacial tension at the time the drop is formed. A lower interfacial tension decreases the energy required to disperse the oil. Since the aging increases the dynamic value of the interfacial tension, the initial drop number decreases for a system whose interface takes longer to age.

Our observations on the drainage of these protein films showed that the time required for the films to drain was about 4 minutes for a film with a radius of 80 μm (drop size of 280 μm). The protein concentration in the aqueous solution was 0.1 wt%. After draining, the final thickness of the stable film was large ( > 1500 Å) and non-homogeneous. Fig. 5 shows the film after it had drained. The contrast in the reflected light from different parts of the film shows that the films are nonuniformly thick.

These films remained stable for over a week. Graham and Phillips (16-17) found that proteins, like BSA, adsorb as multilayers at the oil/water as well as the air/water interfaces. They found the thickness of the multilayers to be close to 250 Å for air/water interface. The thick stable films that we observed are probably stabilized by the interactions between the multilayers of adsorbed protein at the two interfaces of the film.

Emulsifier Films

Interfacial viscosity measurements were made on the solutions containing the emulsifier (EOM) and Fig. 6 shows the ratio of the velocity of the dish to the velocity for four different concentrations of the emulsifier as well as for the pure oil-water system. We see that the particle reaches a constant velocity immediately and, for all the emulsifier concentrations studied as well as the pure system, the velocity is the same. Thus, there is no measurable change in the interfacial viscosity due to the emulsifier.

The interfacial tension isotherm for the emulsifier system is shown in Fig. 7. We found the critical micelle concentration (cmc) to be around $4x10^{-5}$ wt%. This is more than 10000 times lower than the lowest emulsifier concentration used, for the interfacial viscosity measurements as well as for the emulsion stability tests. At these emulsifier concentrations the tension gradients are not likely to exist due to the rapid diffusion from the bulk to the interface. The interfacial tension measurements did show that at these concentrations the equilibrium interfacial tension values were reached rapidly. The slope of the interfacial tension isotherm was used to calculate the area covered by the emulsifier at the interface. This value was used, along with the interfacial area obtained from the emulsion stability tests, to determine the amount of emulsifier depletion from the bulk solution due to interfacial adsorption.

Fig. 8 shows the emulsion stability tests for the emulsifier system. The emulsion stability (characterized by the decrease in the number of drops) decreases with the concentration of the emulsifier. In this case, too, we see that the initial drop number decreases with the decrease in emulsifier concentration. The depletion of emulsifier, due to adsorption at the interface, was calculated and found to be negligible. At surfactant concentrations much higher than the cmc, which is the case for the emulsions that we tested, the interfaces are saturated and tension gradients are minimized. Wasan and co-workers (18,19) found the drainage times to increase with surfactant concentrations for both emulsion and foam films. Manev et al. (18,19) and Nikolov et al. (3,4) found that at high surfactant concentrations foam and emulsion films exhibited step wise transitions in the film thickness, a phenomenon termed as stratification, which inhibited film drainage. According to Nikolov and Wasan (2), stratification, ie. a layer by layer thinning, is due to the presence of ordered structures of colloidal aggregates inside the film. Since the emulsifier concentrations used for the present emulsion system is over 10000 times the cmc, it is highly likely that stabilization is due to micellar structuring in the film. Dynamic light scattering confirmed the presence of micellar aggregates in solutions of the emulsifier and their size distribution is shown in Fig. 9. The mean size of these aggregates is about 118 nm.

In recent studies, Nikolov et al. have investigated the stratification phenomenon, in films formed from bulk solutions, by carrying out systematic experiments with ionic (3,4) as well as nonionic (20) micellar surfactant systems. The stratification phenomenon is explained due to the formation of ordered structures of micelles within the film. As the film thins, the micelles, within, interact via long-range screened electrostatic (in the case of ionic systems) or steric (nonionic micelles) repulsions, due to the restricted volume conditions in the film. Under these conditions the micelles form an "ordered" structure, within the film, thereby reducing the free energy of the system. A step transition in the film thickness occurs when a single layer of micelles is expelled from the film.

Reflected light microscopic studies for the emulsifier stabilized films showed that there are transitions in the film thickness occurring during thinning. Initially, when the films were formed, they were bright (indicating thick films). As the films drained, a grey band (indicating thinning) appeared around the perimeter. This indicates the presence of a dimple, which are common to emulsion films. Fig. 10 shows a part of an emulsion film, from where thinning begins. The grey region around the periphery of the film indicates a thickness that is less than 1500 Å. The black area adjoining the grey area indicates a sharp transition to a lower thickness. Before the films turned black we saw two such transitions. This is similar to the stratification phenomenon observed by Manev et al. (19) in planar emulsion films and Nikolov et al. in foam films (3,20). The completely drained films containing the emulsifier were very thin (black films) and uniform unlike the protein stabilized films.

## CONCLUDING REMARKS

In the present study two main mechanisms were found to stabilize soy bean oil-in-water emulsions. Interfacial rheological properties, such as interfacial viscosity, were found to play an important role in the stabilization of protein stabilized emulsions. The protein adsorbs as multilayers at the interface, which results in extremely high interfacial viscosity. Due to the slow adsorption/diffusion kinetics of proteins, the interface exhibits an aging effect. The aging time of the interface is dependent on the protein concentration. A high interfacial viscosity as well as the aging effect control the stability of these emulsions.

Emulsions containing a food grade nonionic emulsifier (EOM) were found to be stabilized by the mechanism of stratification. The emulsifier concentrations which are of practical interest are higher than the critical micelle concentration. Dynamic light scattering measurements showed that micellar aggregates existed in solutions of the ethoxylated monoglyceride (EOM) and the emulsion films of this system exhibited two sharp transitions in the thickness while draining. This phenomenon of stratification, which leads to an enhanced stability of emulsion and foam systems, is similar to that observed by Wasan and co-workers (3,4,20), which they showed to be caused by the presence of colloid crystal-like structures within the film.

ACKNOWLEDGEMENT

This work has been supported by Kraft General Foods.

LITERATURE CITED

1. Sherman, P., Emulsion Science, Academic Press, New York, (1968).

2. Friberg, S. and Rydhag, L., Kolloid-Z u. Z. Polymere, 244, 233, (1971).

3. Nikolov, A.D. and Wasan, D.T., J. Colloid and Interface Sci., 133(1), 1, (1989).

4. Nikolov, A.D., Kralchevsky, P.A., Ivanov, I.B. and Wasan, D.T., J. Colloid Interface Sci., 133, 13, (1989).

5. Friberg, S., Mandell, L. and Larsson, M., J. Colloid Interface Sci., 29, 155, (1969).

6. Krog, N., Barfod, N.M. and Sanchez, R.M., J. Disp. Sci. and Tech., 10, 483, (1989).

7. Gaonkar, A., personal communication.

8. Mannheimer, R.J. and Schechter, R.S., J. Colloid Sci., 32, 195, (1970).

9. Pintar, A.J., Israel, A.B. and Wasan, D.T., J. Colloid Interface Sci., 37, 52, (1971).

10. Wasan, D.T., Gupta, L. and Vora, M.K., AIChE J., 17, 1287, (1971).

11. Gupta, L. and Wasan, D.T., Ind. Eng. Chem. Fundam., 13, 26, (1974).

12. Menon, V.B. and Wasan, D.T., Sep. Sci. and Technol., 19, 555, (1984).

13. Zapryanov, Z., Malhotra, A.K., Aderangi, N. and Wasan, D.T., Int. J. Multiphase Flow, 9, 105, (1983).

14. Malhotra, A.K. and Wasan, D.T., Chem. Engg. Comm., 55, 95, (1987).

15. Graham, D.E. and Phillips, M.C., J. Colloid Interface Sci., 70, 403, (1979).

16. Graham, D.E. and Phillips, M.C., J. Colloid Interface Sci., 70, 415, (1979).

17. Graham, D.E. and Phillips, M.C., J. Colloid Interface Sci., 70, 427, (1979).

18. Manev, E.D., Sazdanova, S.D. and Wasan, D.T., J. Dispersion Sci. Technol., 3, 435, (1982).

19. Manev, E.D., Sazdanova, S.D. and Wasan, D.T,, J. Dispersion Sci. Technol., 5, 111, (1984).

20. Nikolov, A.D., Wasan, D.T., Kralchevsky, P.A. and Ivanov, I.B., "Ordered Structures in Thinning Micellar Foam and Latex Films", in Ordering and Organization in Ionic Solutions -

Proceedings of Yamada Conference XIX, N.Ise and I. Sogami (Eds.), World Scientific Publishing Co., Singapore (1988).

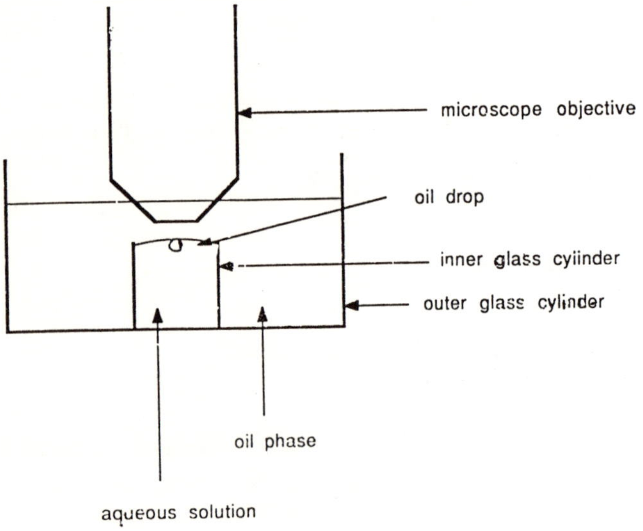


Figure 1. Schematic of the experimental cell used for thin film studies.

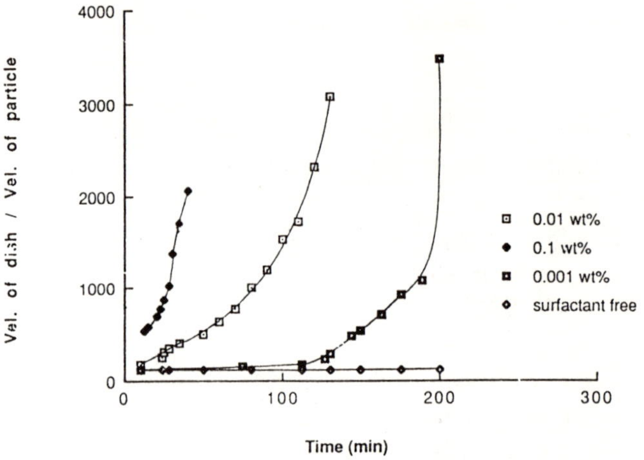


Figure 3. Interfacial viscosity measurements for BSA at the soybean oil/water interface.

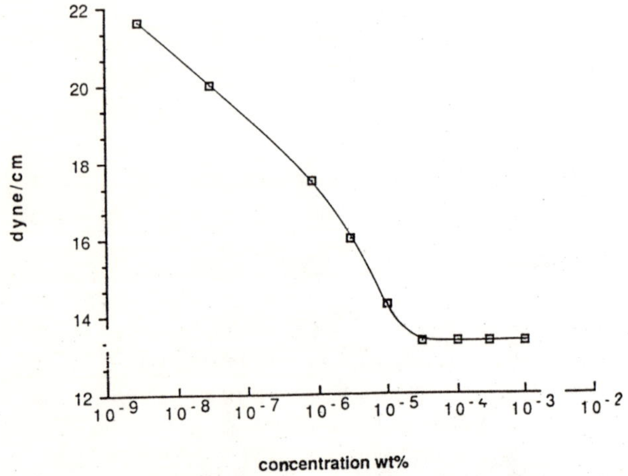


Figure 2. Interfacial tension isotherm for BSA at the soybean oil/water interface.

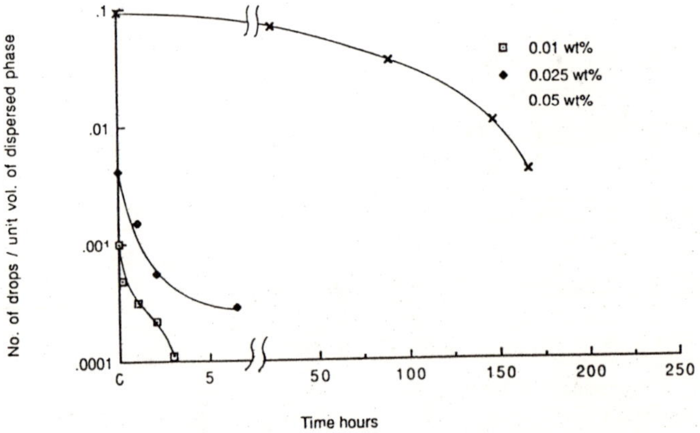


Figure 4. Number of drops plotted as a function of time for three emulsions of soybean oil in water, stabilized by BSA.

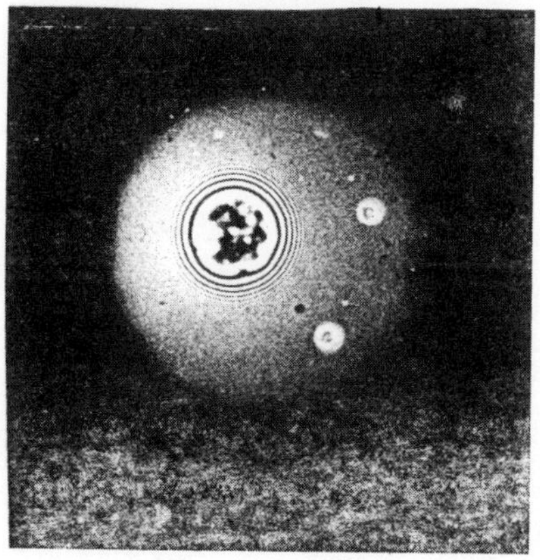

Figure 5. Protein stabilized emulsion film for the BSA-soybean oil system. The film is bounded by the first dark interference ring and has a diameter of 80 $\mu$m. The diameter of the oil drop is about 350 $\mu$m.

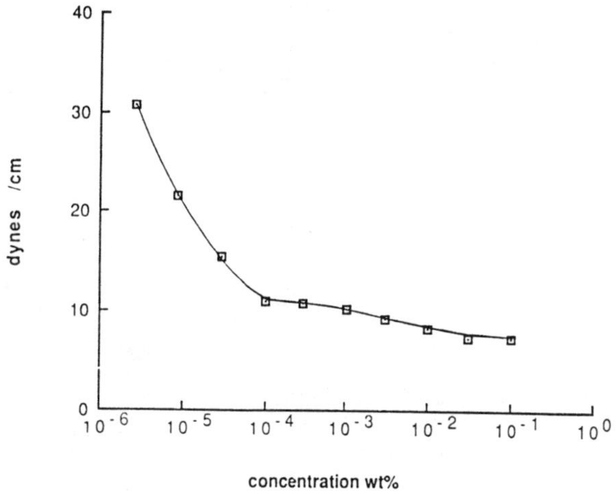


Figure 7. Interfacial tension isotherm for EOM at the soybean oil/water interface.

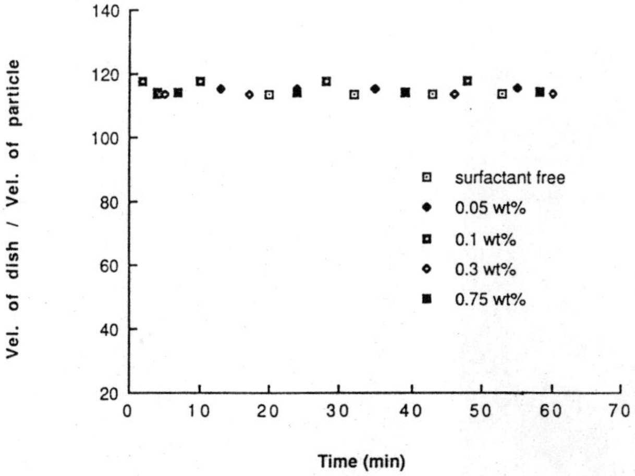


Figure 6.

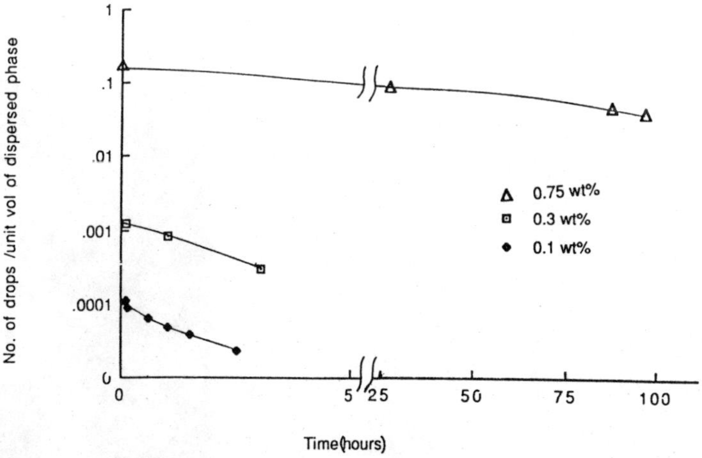


Figure 8. Number of drops plotted as a function of time for three emulsions of soybean oil in water stabilized by EOM.

```
------------------------------------------------------------------------------
Particle size distribution ( nm. ) Graph of Distribution of number
    9.1   -    11.0  |+
   11.0   -    13.3  |+
   13.3   -    16.1  |+
   16.1   -    19.5  |+
   19.5   -    23.7  |+
   23.7   -    28.7  |+
   28.7   -    34.7  |+
   34.7   -    42.1  |+
   42.1   -    51.0  |+
   51.0   -    61.8  |+
   61.8   -    74.8  |-----------------------------------------------+
   74.8   -    90.7  |-----------------------------------------+
   90.7   -   109.8  |--------------------+
  109.8   -   133.1  |-------+
  133.1   -   161.2  |----+
  161.2   -   195.3  |--+
  195.3   -   236.7  |-+
  236.7   -   286.7  |+
  286.7   -   347.4  |-+
  347.4   -   420.9  |-+
  420.9   -   509.9  |+
  509.9   -   617.7  |+
  617.7   -   748.4  |+
  748.4   -   906.7  |+
------------------------------------------------------------------------------
```

Figure 9. Size distribution of the colloidal aggregates present in a 0.1 wt% EOM solution.

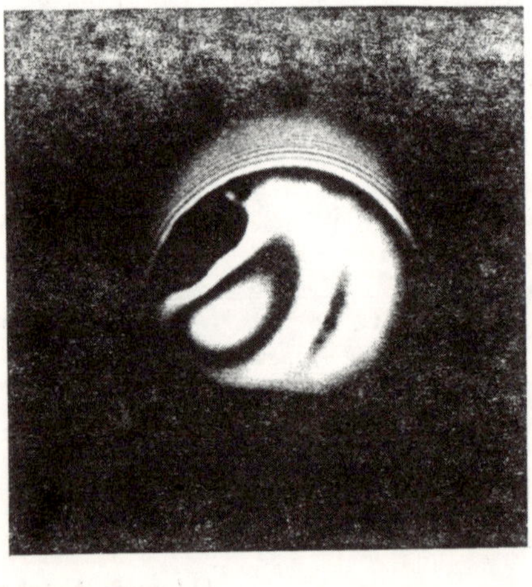

Figure 10. A draining emulsion film which is stabilized by the emulsifier EOM. The film is formed from an oil drop which is about 1 mm in diameter.

# DESIGN AND SELECTION OF EMULSIFIERS IN THE FOOD INDUSTRY

Gerard L. Hasenhuettl ■ Kraft General Foods, Kraft, USA Technology Center, 801 Waukegan Road, Glenview, IL 60025

Emulsifiers are essential ingredients to stabilize emulsion based food products during commercial storage. Their functionality depends on a polar hydrophilic end and a non-polar lipophilic moiety. Emulsifiers are designed by molecular variation of these structures. These substances are regulated by the Food and Drug Administration as either Generally Recognized As Safe or as Regulated Direct Food Additives. Selection of emulsifiers for food systems depends on functionality, cost, secondary effects, and regulatory considerations. After a system has been selected, it may be optimized through statistical experimental design.

## INTRODUCTION

Emulsion based food products were historically prepared using naturally occurring emulsifiers such as egg yolk. However, the revolution in processed foods has spawned a demand for emulsifiers which maintain the stability of emulsion based foods for as much as a year while the product resides in commercial distribution.

In addition, research has demonstrated that surface active agents may produce secondary technical effects in the food product. The secondary properties may be more critical in consumer acceptance of a product than is stability of the emulsion. Examples of these effects are: anti-staling effect of sodium stearoyl lactylate in bread, aeration of cake mix batters by propylene glycol monoesters, and inhibition of crystal formation by polyglycerol esters in salad oil when stored in a refrigerator.

Structurally, emulsifiers are composed of a polar hydrophilic (water loving) group and a non-polar lipophilic (lipid loving) chain. Figure 1 shows a schematic representation of an amphiphilic molecule. Polar groups may be varied such as carboxylate or hydroxyl groups. Lipophilic groups are restricted to saturated or unsaturated fatty acids. Surfactants are commonly classified as anionic, cationic, amphoteric, or nonionic depending on the type of charge carried by the molecule. Most food emulsifiers are nonionic. Their functionality is effected by the dipole orientation and hydrogen bonding capability of the hydroxyl group. Protein and lecithin may be amphoteric or cationic, depending on the pH of the food system.

One classification system for emulsifiers is the HLB (hydrophile-lipophile Balance) of the molecule (1). Low HLB emulsifiers are non-polar and useful for formulation of water in oil (W/O) emulsions. High HLB substances are polar and tend to be used for oil in water (O/W) emulsions. The emulsifier is conventionally dispersed in the continuous phase but exceptions to this rule are frequent. Table I lists HLB values for some typical emulsifiers sold by food ingredient suppliers (2).

## REGULATORY CONSIDERATIONS:

Emulsifiers, as with other food ingredients, are regulated in the United States by the Food and Drug Administration (3). Other countries may have their own regulations concerning food additives. Marked differences may exist between countries. For example, polysorbates are permitted in the U.S. but not in Japan. Ricinoleate esters which are permitted in

Table 1. HLB Values of Some Commonly Used Food Emulsifiers.

| **Emulsifier** | **HLB** |
|---|---|
| Decaglycerol decaoleate | 2.0 |
| Propylene glycol monostearate | 2.2 |
| 42% Monoglyceride | 2.8 |
| 52% Monoglyceride | 3.3 |
| Distilled Monoglyceride | 3.5 |
| Sorbitan monostearate | 4.7 |
| Triglycerol monoester | 7.2 |
| Polysorbate 65 | 10.5 |
| Octaglycerol monoester | 13.0 |
| Ethoxylated Monoglyceride | 13.0 |
| Polysorbate 60 | 14.9 |
| Polysorbate 80 | 15.9 |

Japan are not approved in the U.S. the FDA classifies emulsifiers into two major groups: 1) Generally Recognized as Safe (GRAS). These additives have a history of safe use and may be used in any non-standardized food product at any level necessary to obtain the desired technical effect. GRAS emulsifiers are listed in Table II. 2) Regulated Direct Food Additives. These substances may be regulated similarly to GRAS status, such as propylene glycol esters. More often, they are strictly regulated as to method of manufacture, analytical constants, types of foods in which they may be used, and maximum permissible concentrations. Table III lists some commonly used direct food additives.

## EMULSIFIER PRODUCTION AND APPLICATIONS:

### Lecithin

Lecithin is a natural phospholipid emulsifier. Its structure is based on glycerine which is esterified with two fatty acids and a phosphoric acid. the phosphate is bonded to an organic base, usually choline or ethanolamine or a carbohydrate such as inositol. Although egg lecithin is commonly purified and used in pharmaceuticals, the most common food ingredient is derived from soybeans. Soy oil can be degummed and commercial lecithin contains an appreciable concentration of residual oil. Lecithin can be further purified by crystallization from an organic solvent (4). Of course additional purification also drives up the cost of the ingredient. A common use for lecithin is in chocolate. Here the term emulsifier is misapplied since there is no aqueous phase. The secondary effect of viscosity reduction is the most important feature in this application. Lecithin is also used as a co-emulsifier in margarines and low fat spreads.

### Monoglycerides

Monoglycerides are composed of a glycerine molecule esterified with a single fatty acid. These ingredients are GRAS food additives in the United States. Commercially, monoglycerides may be produced by direct esterification with fatty acids or by interesterification with fats (5), the latter being less expensive. Figure 2 illustrates these methods. Interesterification may be carried out in a batch or continuous (6) ester reactor. Monoglycerides may be sold as they are produced from the reactor (mono/diglycerides, 40-50% mono) or they may be purified by short path distillation to monoglyceride contents of >90%. Monoglycerides are used widely in food products. Some control over properties may be obtained by variation of the fatty acid side chain. For example, semi-saturated monoglycerides may be useful aerating agents.

### Propylene Glycol Esters (PGMEs):

Propylene glycol esters are identical to monoglycerides except that they are missing one terminal hydroxyl group. Although they are regulated Direct Food Additives, they may be used in virtually any non-standardized food product. Like monoglycerides, they may be prepared by direct or interesterification (7). Their primary use in processed foods is for aeration in cake mix batters. The fatty acid moieties in these products are almost always saturated (palmitic and stearic).

### Lactylated Monoglycerides and PGMEs:

Monoglycerides and propylene glycol fatty acid esters may be further reacted with lactic acid resulting in a modification of the polar head group of the emulsifier. They are direct food additives which are regulated for their method of manufacture, analytical constants, usage levels, and the types of food in which they may be used. These compounds are useful for emulsion stabilization and aeration of whipped toppings. As with PGMEs the fatty acid residues in these compounds are invariably saturated (8).

Other Derivatives of Monoglycerides:

Monoglycerides may be ethoxylated (9), succinylated (10), acetlyated (11), citrated (12), phosphorylated (13), or reacted with diacetyltartaric acid (14) to form modified derivatives. Phosphated monoglycerides and diacetyltartaric acid esters (DATEM) appear on the FDA's GRAS list. The other materials are Direct Food Additives. Uses of the derivatives vary widely but are usually noted for their secondary functional effects rather than for emulsification. For example, succinylated monoglycerides are good dough strengtheners in baked goods.

Sorbitan Esters and Polysorbates:

Sorbitan esters are produced by the simultaneous esterification and cyclization of sorbitol with a fatty acid (15). Polysorbates are produced by further reaction of sorbitan esters with ethylene oxide. Sorbitan monostearate is a strictly controlled Direct Food Additive. Interestingly sorbitan monooleate is not an approved food additive in the U.S. but in the polysorbate derived form it is a Direct Food Additive. Sorbitan esters and polysorbate provide an interesting example of synergy. When used in combination, they often provide better functionality then either individual emulsifier. Common applications are whipped toppings, coffee whiteners, and salad dressings.

Polyglycerol esters (PGEs):

The class of polyglycerol esters constitutes a versatile family of emulsifiers. Subtle functional behaviors may be controlled by the chain length of the polyglycerol chain and the saturation/unsaturation of the fatty acid (16). The HLB value may be varied through a wide range by controlling the ratio of the fatty acid to polyglycerol. Although they are Direct Food Additives, the PGEs have wide approval in nonstandardized foods. They may be used in any concentration. However, they are generally self-limiting due to their distinctive bitter aftertaste when levels are too high. Some progress has been made in recent years toward lower levels of off flavors and odors by some manufacturers. Applications of the PGEs are quite diverse. They find use as aerating agents, crystal modifiers, and anti-spattering agents.

Sucrose Esters:

As a class, sucrose esters also display some versatility. Composition of the fatty acid and its ratio to sucrose may be varied to obtain desired functionality. In practice, only the mono-, di-, and triesters are used as emulsifiers (17). Tetraesters and above constitute the family of components known as sucrose polyesters. These compounds are not digestible and are utilized as fat substitutes. The emulsifiers are generally prepared by interesterification of sucrose with fatty acid esters. In order to obtain sufficient solubility for reaction at low temperatures, a polar solvent or high level of soap is required. A solvent crystallization step is then necessary to remove excess soap. Solvent removal is essential to ensure the safety of the product. This process tends to make the emulsifier costly. Although widely used in Japan, the product is only approved for bakery products, frozen desserts, and protective fruit coatings in the U.S.

Sodium and Calcium Stearoyl Lactylate:

These compounds represent the most widely used anionic emulsifiers widely used as ingredients in the food industry. They are prepared by esterification of stearic and lactic acids concurrent with neutralization with sodium or calcium hydroxide. These products are Direct Food Additives and are controlled strictly as to composition, analytical constants, levels, and types of foods in which they may be used. Their most common application is in texture control of baked products (e.g. retardation of staling in bread).

EMULSIFIER SELECTION:

The selection of emulsifiers for formulated food products has traditionally been an art. Experienced practitioners could design a successful emulsifier system with a limited number of experiments. However, the novice in this field faces a bewildering scene of possibilities, regulations, and technical problems. To add to this complexity, it is common practice to use emulsifiers in combination to obtain a balance of desired effects. For example, a cake mix needs to aerate to a low specific gravity, produce an even grained structure, and retain a relatively high level of moisture through the baking process.

Suppliers have been and continue to be an excellent source of information concerning their products. However, due to recent mergers and acquisitions, U.S. suppliers may be experiencing shortages in R&D staffs. if an emulsifier can be used "off the shelf" a supplier can usually supply it. However, if breakthrough technology is required for a radically different new food product, the supplier may not be capable to provide support.

The following guidelines are therefore suggested for approaching the problem of emulsifier selection:

REGULATORY QUESTIONS:

Does the Product Have a Standard of Identity?

Some food products such as margarine, French dressing, peanut butter, and others are defined by the FDA to have a specific composition. This often limits which emulsifiers (if any) may be used and at what level. However, changes in a Standard of Identity may be petitioned. For example, the standard for margarine was recently changed to allow the use of any food approved emulsifier at a concentration sufficient to obtain the desired effect. Also, the food product may avoid the standard by simply not using the standard name.

Does the Food Product need to be "All Natural"?

This consideration may be a serious limitation to emulsifier use. If the "all natural" designation is necessary, the developer is limited to substances which are naturally occurring and not artificially obtained. For example, monoglycerides occur in nature but when they are produced commercially, they are not natural. Lecithin and proteins are the only practical natural food emulsifiers.

Is the Proposed Emulsifier GRAS?

If the emulsifier under consideration is on the FDA GRAS list, it may be used in food products where it is not precluded by a Standard of Identity. Some Direct Food Additives may also have broad approval. However, as previously mentioned, some emulsifiers are strictly regulated and the food product being developed may not be an approved use for the particular emulsifier. A petition may be filed for expanded use but the procedure may be costly and time consuming.

Does the Product Need to be Kosher?

Several Rabbinical Councils certify food products and ingredients as conforming to Kosher dietary requirements. Most food emulsifiers can be obtained in Kosher form. One notable problem may arise with sucrose esters which are specified by the FDA to be produced from tallow fatty acids. One must also be certain to ascertain whether the particular council certifying the ingredient is acceptable to the organization which certifies the finished food product.

TECHNICAL QUESTIONS:

Is the Food an Oil in Water or Water in Oil System?

In commercial practice, the emulsifier is most often added to the continuous phase. Hydrophilic emulsifiers (high HLB) disperse more readily in water. Lipophilic (Low HLB) emulsifiers are more oil soluble. However, in some cases, dispersion of the emulsifier in the dispersed phase may lead to smaller droplet sizes and more stable emulsions.

What are the Desirable Secondary Effects?

As previously mentioned, functions other than emulsion stability are important for consumer acceptance of a product. For example, a whipped topping needs aeration and foam stability in addition to emulsification of fat. It is not uncommon to use two or more emulsifiers to obtain a combination of technical effects.

What is the best optimization procedure?

The answer to this question is much too large for the scope of this article. Once an emulsifier has been selected, an effort must be made to obtain the most efficient and informative experimental design to obtain the optimal level of each emulsifier. For example, Figure 3 shows a simple factorial design for a mythical product containing three emulsifiers. Acceptance of this mythical product by the consumer was found to depend on two attributes; Light texture as a result of aeration and viscosity which produced a creamy texture. In addition, emulsion stability was required to maintain the integrity of the product. Separation as measured is the inverse of stability. A cursory examination of the data shows that emulsifier C is the major contributor to stability. Emulsifier A builds viscosity and B allows the product to be whipped to a low specific gravity. This simple factorial design can now serve as a basis for selecting a three dimensional space for a second design to further optimize the concentrations of A, B, and C. For a more complete treatment of experimental design, the reader is referred to a text on experimental design such as Box, et. al. (18).

Although emulsifier selection can be viewed as a complex process, the competitive realities of food product development will require a streamlined selection process. This will be particularly true as foods evolve to meet consumer expectations of lower fat, more nutritious and more convenient products.

LITERATURE CITED:

1. a) Griffin, W. C., J. Soc. Cosmet. Chem. 1, 311, 1949. b) Becher, P., Encyclopedia of Emulsion Technology, Vol. 2, New York, Marcel Dekker, Inc., 1985, pp. 425-512.

2. Durkee Industrial Foods Bulletin, "Santone Polyglycerol Esters", IND-165, 1982.

3. Federal Register, Code of Federal Regulations, Vol. 21, Parts 170-199, Washington, DC, Government Printing Office, 1988.

4. Szuhaj, B., Lecithins: Source, Manufacture, Uses, Champaign, IL, American Oil Chemists Society, 1985.

5. a) Christenson, C.N., U.S. Pat. 2,002,493 (1936). b) Edeler, A., and A.S. Richardson, U.S. Pats. 2,206,167 and 2,206,168 (1940). c) Goldsmith, H.A., Chem. Rev., 33, 257, 1943.

6. Allen. R.R. and R.L. Campbell, U.S. Pat. 3,313,834 (1967).

7. Kuhrt, N.H., et. a., U.S. Pat. 3,034,857 (1962).

8. Woods, G.E., U.S. Pat. 3,012,047 (1961).

9. Norris, M.E., U.S. Pat. 3.672,458 (1972).

10. Freund, E.H., U.S. Pats. 3,293,272 (1966) and 3,370,958 (1968).

11. a) Bauer, F.V., J. Am. Oil Chem. Soc., 31, 196-9, 1954. b) Feuge, R.O., et. al., U.S. Pat. 2,745,749 (1956)/

12. a) Radlove, S., U.S. Pat. 3,227,550 (1966). b) Goldschmidt, T.H., Belg. Pat. 834950 (1976).

13. a) (Cawley, J.D. and M.P. O'Grady, U.S. Pat. 3,423,440 (1960). b) Meguro, T., et. al., Jap. Pats. 50/25458 (1975) and 51/47692 (1976).

14. Gladstone, M.M., U.S. Pat 2,938,027 (1960).

15. a)Griffin, R.C., U.S. Pat. 2,380,166 (1945). b) Stockburger, G., U.S. Pat. 4,297,290 (1980).

16. a) Babayan, V., U.S. Pat. 3,637,774 (1972). b) McIntyre, R., J. Am. Oil Chem. Soc., 56: 835A-40A, (1979). c) Seiden, P., and J.B. Martin, U.S. Pat. 3,968,169 (1976).

17. a) Osipow, L., et. al., J. Am. Oil Chem. Soc., 34: 185-8, 1970. b) Kinami, K., et. al., Jap. Pat. 62/129294 (1987). c) Feuge, R.O., et. al., J. Am. Oil Chem. Soc., 47: 56-60, 1970. d) Osipow, L., and W. Rosenblatt, J. Am. Oil Chem. Soc., 44: 307-9, 1967. e) Yamagishi, F., et. al., Ger. Pat. DE 2,051,766 (1971). f) Yamamoto, T., and K. Kinami, Jap. Pat. 61/212593 (1986).

18. Box, G.E.P., W.G. Hunter, and J.S. Hunter, Statistics for Experimenters, New York, John Wiley & Sons, 1978.

Table 2. Emulsifiers generally recognized as safe.

| EMULSIFIER | 21 CFR NO. | FUNCTIONALITIES | TYPICAL APPLICATIONS: |
|---|---|---|---|
| 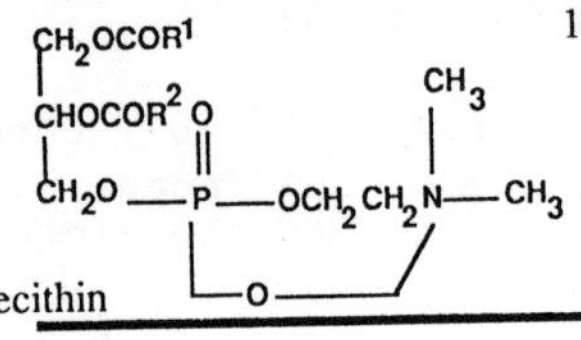 Lecithin | 184.1400 | Co-Emulsifier, Viscosity Reducer | Margarines, Chocolate Products |
| 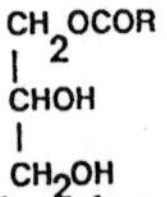 Monoglycerides | 182.4505 | Emulsification, Aeration, Crystal Stabilization | Margarines/Spreads, Whipped Toppings, Peanut Butter Stabilizers |
| 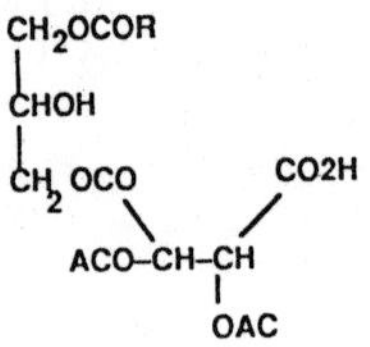 ltartaric Esters of Monoglycerides (DATEM) | 182.4101 | Emulsifier, Film Former | Dough Strengthener |
| 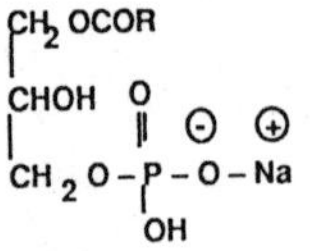 Phosphated Monoglyceride (Monosodium Salt) | 182.4521 | Amionic Emulsifier | Emulsifier, Co-emulsifier |

Table 3. Emulsifiers which are direct food additives.

| EMULSIFIER | 21 CFR NO. | FUNCTIONALITIES | TYPICAL APPLICATIONS: |
|---|---|---|---|
| 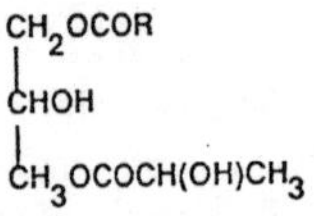 Lactylated Monoglycerides | 172.852 | Aeration, Viscosity Control | Cakes, Whipped Toppings |
| 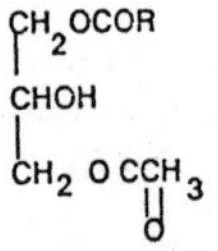 Acetylated Monoglyceride | 172.828 | Film Former | Edible Moisture Barrier for coating fruits, nuts, pizza, etc. |
| 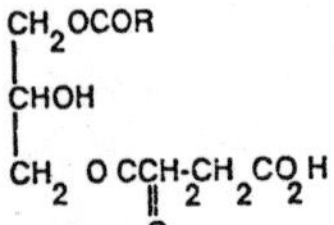 Succinylated Monoglyceride (SMG) | 172.830 | Emulsifier, Complex Former | Emulsifier, Dough Conditioner in baked products |
| 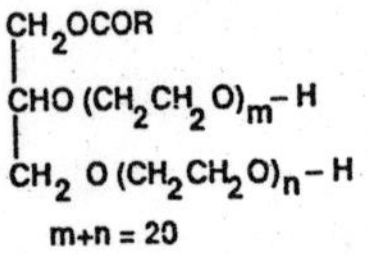 Ethoxylated Monoglyceride | 172.834 | Emulsifier, Aerator, Complex Former | Coffee Whiteners, whipped toppings, cakes, baked products |

Table 3. (continued from Page 40).

| | | | |
|---|---|---|---|
| 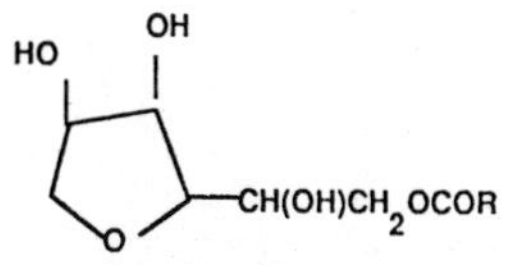 Sorbitan Esters | 172.842 | Emulsifier / Co-Emulsifier | Cake Icings, Coffee Whiteners, Confectionery Coatings |
| 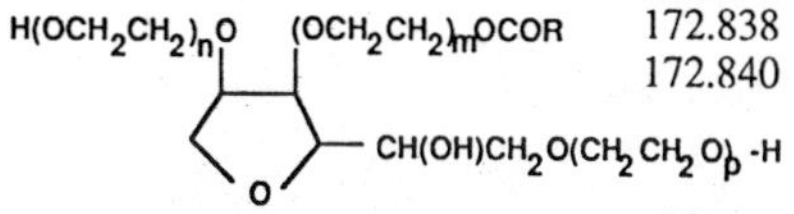 Polysorbates $m + n + p = 20$ | 172.836<br>172.838<br>172.840 | Emulsifier, solubilizer | Whipped toppings, coffee whiteners, flavor carriers |
| 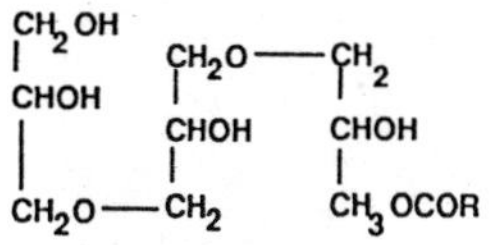 Polyglycerol Esters | 172.854 | Emulsification, Aeration, Crystal Modification | Beverage Clouding Agents, Icings, Whipped Toppings, Salad Oils |
| 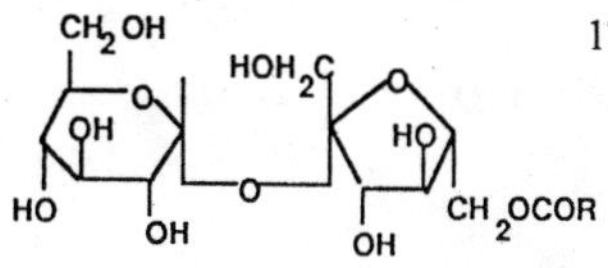 Sucrose Esters | 172.859 | Emulsifier, Film Former, Moisture Barrier | Baked Goods, Frozen Desserts, Protective Coatings, on Fruit |
| 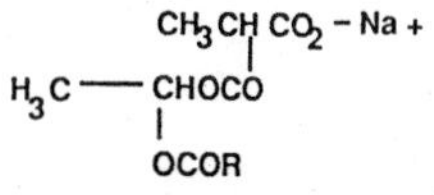 Sodium Stearoyl Lactylate | 172.846 | Emulsification, Freeze-Thaw Stabilization, Starch Complexation | Coffee Whiteners, Bread, Dehydrated Potatoes, Chip Dips |
| 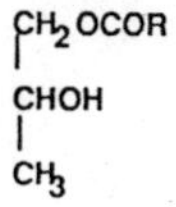 Propylene Glycol Esters | 172.856 | Aeration, increase in interfacial viscosity / elasticity | Cake mixes, Whipped toppings |

Food Emulsifier Structure

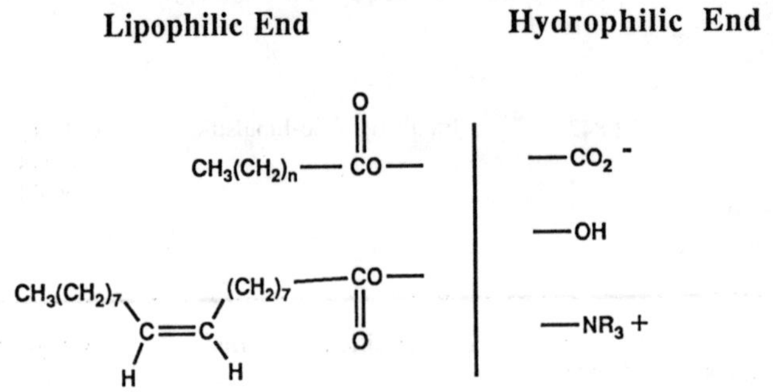


**Emulsifier Types:**

**Cationic (Positive Charge)**

**Anionic (Negative Charge)**

**Amphoteric (Positive and Negative Charges)**

**Nonionic (Dipole Without Formal Charge**

Figure 1. Molecular structure of food emulsifiers.

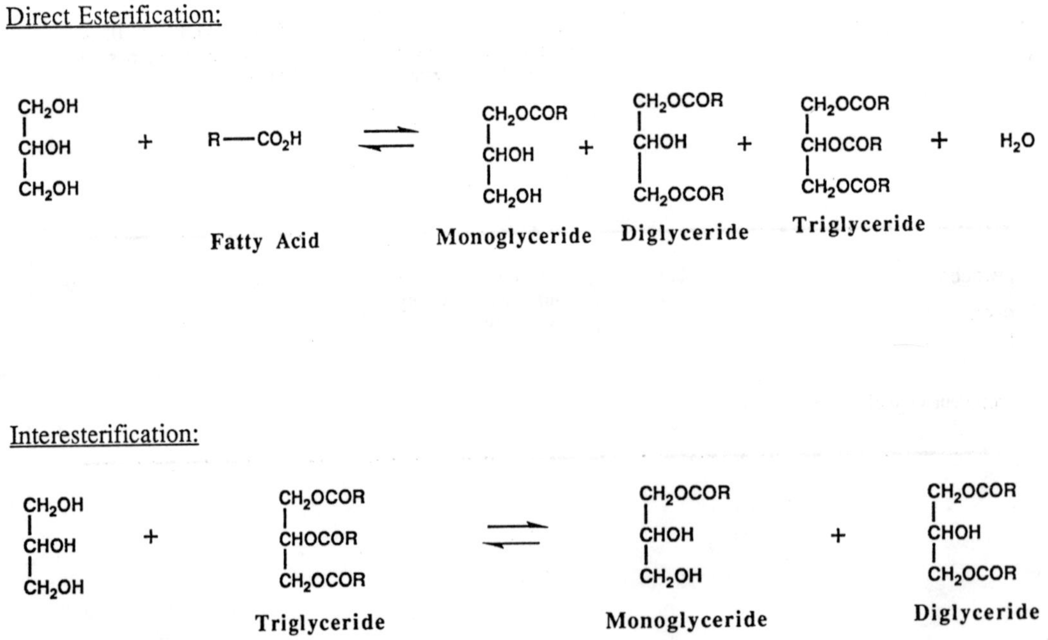


Figure 2. Preparation of monoglycerides by direct and interesterification.

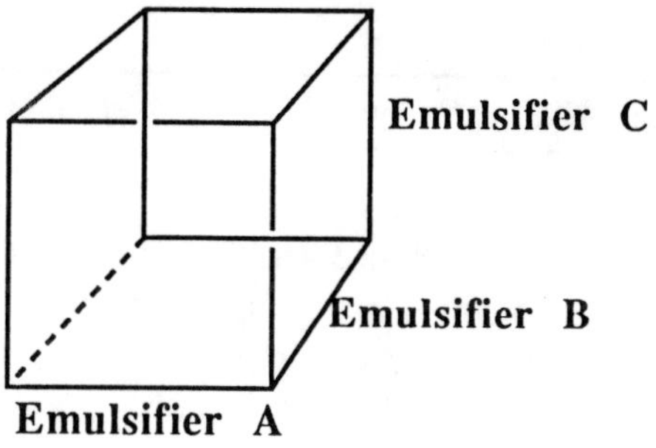


| EMULSIFIER A | B | C | Viscosity (cps) | Sp. Grav. (g/cc) | Separation (g) |
|---|---|---|---|---|---|
| - | - | - | 270 | 0.982 | 89.9 |
| + | - | - | 960 | 0.715 | 92.3 |
| - | + | - | 1025 | 0.440 | 84.3 |
| - | - | + | 250 | 0.930 | 3.1 |
| + | + | - | 6280 | 0.590 | 24.8 |
| + | - | + | 775 | 0.695 | 2.2 |
| - | + | + | 1180 | 0.515 | 1.9 |
| + | + | + | 5860 | 0.575 | 0.4 |

| | | |
|---|---|---|
| **Emulsifier A** | **- = 0.1 %** | **+ = 1.5 %** |
| **Emulsifier B** | **- = 0.05 %** | **+ = 1.0 %** |
| **Emulsifier C** | **- = 0.01 %** | **+ = 0.5 %** |

Figure 3. Simple factorial design for three emulsifiers.

# EMULSION PROPERTIES OF MARGARINES AND LOWFAT SPREADS

Rajendra P. Borwankar and Gregory S. Buliga ■ Kraft General Foods, 801 Waukegan Road, Glenview, IL 60025

Margarines and lowfat spreads are examples of water-in-oil food emulsions. The oil blend used in these systems usually contains partially hydrogenated vegetable oil. The texture of margarines and spreads is governed by the type and amount of crystals in the oil phase. The aqueous phase usually contains thickeners and/or dairy proteins. These additives improve the flavor release characteristics of these systems. Simple tests used as a guide in the development of lowfat spreads as well as processing issues associated with their production are discussed.

Margarine was invented over 120 years ago by a French chemist, Hippolyte Mege Mouries, in response to the prize announced by the French government for a suitable butter substitute, which would be cheaper and keep better than butter (1). Mouries won the prize and obtained a patent for a butter substitute derived from tallow. In the beginning, margarine carried a negative connotation as a cheap butter substitute. The FDA promulgated a standard of identity for margarine in 1941, giving the product the status of a food substance in its own right. Closely related to margarines are the tablespreads, sometimes called spreads or lowfat spreads. FDA standards of identity require margarine to contain a minimum of 80% fat. Spreads are products with less than 80% fat, and there are no standards of identity for spreads. Diet margarine is a term reserved for spreads with 40% fat.

Margarines and tablespreads are water-in-oil emulsions. The aqueous phase consists of water, salt and preservatives. In many instances, thickeners or milk proteins are also used. The oil phase consists of partially hydrogenated vegetable oil, or occasionally, animal fat. Lecithin and emulsifiers are also usually added to the oil phase. The color used (yellow) is usually β-carotene, which has vitamin A activity. Other vitamins, especially D, may also be used. Artificial flavors are also used. These products come in either stick form or a soft form. Very limited amounts are sold as liquid products.

Margarines and tablespreads (and, of course, butter) represent the most if not the only significant application of the w/o emulsions in the food industry. Water-in-oil emulsions are not as common as the o/w emulsions and have not been studied as well either. Lowering the fat in spreads poses a challenge to the product developer that can be addressed only with a knowledge of emulsion science and technology. In this paper we attempt to highlight the emulsion properties of these products and how we have used the teachings and methods of emulsion science to understand the functionalities of the various ingredients. The application of these learnings has helped us contribute significantly to new product development.

## PROCESSING AND FORMULATION ASPECTS OF MARGARINES AND SPREADS

As shown in Figure 1, the margarine or tablespreads production process involves oil and aqueous phase subprocessing followed by the emulsification process. Emulsification is carried out at about 45-50°C when all the oil is liquid. The critical step subsequent to emulsification is the chilling step. This step is necessary to partially crystallize the oil phase. The lipids exhibit polymorphism, and the desirable crystal form is the β′ form. This form is obtained by

Paper presented at the AIChE Annual Meeting, November 5-10, 1989 in San Francisco.

rapidly cooling the oil. The product is cooled quickly to about 10°C using a scraped surface heat exchanger which facilitates the formation of β′ crystals. The primary purpose of the chilling step is to form many nucleation sites for crystallization to proceed. In the case of stick products, the scraped surface heat exchanger is followed with a crystallization tube to continue the crystallization, so that the product can be formed into sticks. The crystallization proceeds for several hours in the tub after filling.

The first margarine was made with tallow, but over the years the use of vegetable oils has almost completely dominated the margarine and spread industry (2). With growing concern over cholesterol, margarine from vegetable oils has come to be regarded as a healthy substitute for butter, which is a complete turnaround from its earlier image as a cheap butter substitute. The vegetable oil used is usually a blend of partially hydrogenated and non-hydrogenated oils and is tailored to have the desired solid fat index at designated temperatures. Solid fat index (SFI) is a measure of solid fat at the given temperature and can be measured by dilatometry (3). DSC or NMR techniques also provide similar measurements (3). The typical SFI values for the oil blends used in the three type of products are shown in Figure 2 as a function of temperature. All oil blends melt almost completely at about body temperature. The blend used in sticks has the largest amount of crystals at all lower temperatures, and that used for liquid products has a low and almost constant SFI.

## Stick margarines

Stick margarines contain approximately 80% (by weight) oil phase and the remainder is aqueous phase. No thickeners or emulsifiers need be used, and the large number of β′ crystals in the fat phase form a network that stabilizes the emulsion by physically entrapping the water droplets. Milk proteins are added to the aqueous phase to enhance the flavor. Lecithin is added to the oil phase as an anti-sticking and an anti-spattering agent[1]. Lecithin reduces the size of the water droplets because it has a high interfacial activity. The oil-water interfacial tension in the absence of any added emulsifiers[2] is 18-30 mN/m. Lecithin reduces the interfacial tension to below 1 mN/m as shown in Figure 3. The volume average water droplet size in margarines is around 5 µm.

## Soft and liquid margarines

In the case of soft and liquid margarines, the emulsion cannot be maintained by crystal structure alone because there are fewer crystals. Hence, emulsifiers, such as mono- and diglycerides, are used. Here the active component is monoglyceride, but the mixture is cheaper and is therefore used. Lecithin is used for the same reason as stated above.

## Lowfat spreads

Lowfat spreads are the fastest growing category. Spreads from 20-72% fat are on the market. The lowfat products have a healthy image, but as the amount of fat is reduced and that of the aqueous phase is increased, there is a greater need for stabilizing the w/o emulsion. Milk proteins cannot be used because the proteins are interfacially active and tend to stabilize o/w emulsions. Thus, they have a destabilizing effect on the w/o emulsions. In addition to formulating out the milk proteins, one needs to use distilled monoglycerides in the place of mono- and diglycerides for greater emulsion stability. Polyglycerol esters or propylene glycol esters may also be used. Monoglyceride has less interfacial activity than lecithin and even in the presence of the monoglyceride,

1. Margarine is often used for frying purposes. When a pat of margarine is heated, the fat melts and the water droplets become free to move around. The water droplets go to the bottom of the pan where the water evaporates causing spattering. Lecithin is used to reduce the size of the water droplets so each burst is relatively small and spattering is reduced.

2. The oil usually contains native surface active agents which lower the interfacial tension. These are mainly monoglycerides and the high value of 30 mN/m reported above is realized only if the oil is purified (4) to remove these natural surface active agents. Otherwise the behavior is time-dependent with interfacial tension decreasing slowly to about 20 mN/m.

lecithin can further reduce the interfacial tension (Figure 4) indicating a possible competition between the two. The possible competitive adsorption between lecithin and monoglyceride is not well documented and hence poorly understood. Lecithin is also alleged to contribute to organoleptic properties of the product. Finally, in the case of lowfat spreads, aqueous phase thickeners are often used to improve the flavor release for reasons proposed below.

## ROLE OF AQUEOUS PHASE THICKENERS IN LOWFAT SPREADS

Aqueous phase thickeners such as xanthan gum tend to increase the average water droplet size in the product. The effects of larger drop size are two-fold. On the one hand, it makes the spread melt quicker in the mouth and, on the other, larger drops give a larger burst of salt and other water soluble flavors as the product is manipulated in the mouth. We have used water droplet size and meltability index measurements in our work.

Water droplet size measurements on spreads were conducted using laser light scattering (Microtrac II Standard Range Analyzer from Leeds and Northrup) by diluting the sample in hexane containing emulsifiers. The presence of emulsifiers prevents the aggregration of water droplets during the light scattering experiment, which allows an accurate and reproducible measurement of the mean water droplet size. This procedure was shown to be in agreement with the second technique involving droplet size measurements by image analysis using light microscopy.

Meltability of the spreads was determined using a Mettler drop point device. Here, the material is loaded into little cups that have a hole at the bottom. The sample is heated at a constant rate of 1°C/min, and the temperature at which the material becomes fluid enough to drip is called the drop point. It is important to realize that the Mettler drop point is a rheological rather than a thermal concept. It is the temperature at which the sample can flow under its own weight. We have seen it loosely correlate with melting in the mouth as measured by sensory measurements.

Table 1 shows the effect of xanthan gum on the average water droplet sizes and Mettler drop points of 40% fat spreads. It is seen that the addition of xanthan gum increases the average droplet size and reduces the Mettler drop point substantially.

The significant differences in average water droplet sizes in lowfat spreads with and without xanthan gum lead to dramatically different textures of the emulsions during the making of these spreads. The 40% fat emulsions made with xanthan are considerably less viscous than their counterparts made without the gum. This is in agreement with our knowledge of the dependence of the viscosity of the emulsions on the average droplet size (5). In the case of the final product, we see the effect of the droplet size on the rheology reflected in the measured Mettler drop point. Note that the fat crystals in the spreads melt around body temperature in both cases. Clearly, then, the drop points are controlled by emulsion rheology and not by the melting properties of the fat phase.

Our mechanistic understanding of the effect of xanthan gum in increasing drop sizes in lowfat spreads is not complete. We do know that the effect of xanthan gum on the interfacial tension between the aqueous and oleic phases in these systems is negligible. The determination of droplet sizes, during emulsion formation, is governed by breakage and coalescence, whereby the drops are progressively broken down to smaller size until a dynamic equilibrium is reached. In our opinion, the addition of hydrocolloids to the aqueous phase leads to larger droplet sizes primarily because they increase the aqueous phase viscosity and hence the resistance to deformation, thus increasing the difficulty of breaking down the larger drops[3].

---

3. Our earlier hypothesis for the use of thickeners in the aqueous phase was that their addition brings the aqueous phase viscosity closer to that of the oil phase, thus facilitating emulsion formation by favoring drop breakage. See Hinze (6) for the relationship between the ease of breakage of a drop and the ratio of viscosities of the dispersed and continuous phase for drop breakage under viscous flow. Shinnar (7) has suggested that, for large kinematic viscosity of the continuous phase and small interfacial tensions, the conditions seemingly applicable here, even in turbulent flow fields the stresses due to viscous shear within an eddy are dominant. Then, the above arguments regarding the effect of the viscosity ratio would apply to turbulent drop breakage occuring in our process. Clearly, this line of argument contradicts the observations of the water droplet size mentioned above.

As a consequence of these changes in droplet size and meltability, the flavor release is quicker in spreads containing a thickener. hence explaining their widespread use.

Gelatin is often used in lowfat spreads in the place of xanthan gum. We believe that its usage stems from a different line of reasoning. Gelatin is capable of forming gels that melt around body temperature. At high temperatures of processing, the aqueous phase viscosity is relatively unaffected by the addition of gelatin. However, under storage conditions the water droplets exist in a gelled state and as a result are stabilized against coalescence, because the gelled droplets, even after contact, cannot flow together to form a larger drop, but rather stay as a doublet. Thus, it is apparent that the role of gelatin is to increase the product stability. Also, since the gel melts around body temperature, the texture and meltability of the spread in the mouth is unaffected by the addition of gelatin.

**FUTURE CHALLENGES**

We will now turn our attention to the future challenges in the margarine and tablespreads category. We will particularly focus our attention on two major consumer trends that are likely to influence the new products coming out on the market. Both trends are nutritionally driven and are likely to pose major technological hurdles. The consumer in the future is more likely to demand products with low saturated fat content[4]. This is driven by the growing consumer awareness of the detrimental effects of saturated fats on the serum cholesterol levels. This consumer trend has already had its biggest impact on the consumption of tropical oils (8).

The second trend is toward lowering dietary fat intake which will create an increased demand for lowfat products and also create a new demand for the ultra lowfat products. The tonnage in the margarine category in the U.S.A. has been flat or slightly declining in the last few years but lowfat spreads represents a segment that is showing growth (9). The same trend seems to hold true worldwide as well (10).

**Effect of Saturated Fat Content**

As mentioned earlier, varying the level of saturated fat affects the degree of crystallinity of the oil phase. We made 40% fat spreads using oil phases of varying crystallinity obtained by blending liquid soybean oil and stick margarine oil with soft margarine oil. From figure 2 we know that these three oils have different SFI values at the various temperatures below 37°C.

The texture of the spreads made with the different hardness oil blends was visibly very different, increasing in softness in the same direction as the oil itself. This is illustrated by the rheological measurements on these spreads as seen in Figure 5, where G*, which is a measure of firmness, is plotted against the degree of crystallinity. We measured the crystallinity in the oil phase of the low fat spreads by differential scanning calorimetry (Perkin Elmer DSC-7). The enthalpy of melting measured from 7.2°C was taken as a measure of the crystallinity or total solids in the oil phase. The higher the enthalpy of melting in the spreads, the higher the crystallinity. The firmness (texture) of the low fat spreads was measured using a Rheometrics Mechanical Spectrometer from strain sweep experiments at 10°C. G′ and G′′, the storage and loss moduli, were used to calculate G* which was used as a relative measure of the firmness of each spread. Here, $G^*$ is seen to increase with the degree of crystallinity of the oil.

As a consequence of the increased firmness, the stability of the spread is expected to increase with increasing crystallinity in the oil phase. A centrifugation procedure was used to measure the stability of the water in oil emulsions. This technique measures the total amount of oil and water separated from the emulsion after a centrifugal force has been applied. The higher amount of water and/or oil (water, especially) released corresponds to a less stable emulsion. It must be noted that emulsion stability as measured by any accelerated test such as this one only reflects the state of emulsion at that time and may not have any value in predicting the shelf stability of the spreads (11). Figure 6 shows that the stability as measured in the

---

4. Saturated fats are the high melting components in the oil phase and lowering saturated fat levels results in a lower degree of crystallinity.

accelerated test indeed decreases with decreasing crystallinity of the oil phase. This is because the greater amount of crystals in the harder oil blends can hold more liquid oil trapped within their network; the higher firmness of the lipid matrix reduces the sedimentation rates of the water droplets during centrifugation, thereby leading to lower amounts of water (and oil) being separated.

Finally, Figure 7 shows a plot of the Mettler drop point of the spread as a function of the degree of crystallinity of the oil. It is seen that the Mettler drop point of the product is not affected by the degree of crystallinity. These oil blends represent different degrees of hydrogenation of soybean oil and, hence, the melting characteristics of the oils themselves do vary with the degree of crystallinity: the higher the degree of crystallinity in the oil, the higher would be its Mettler drop point. The Mettler drop points of the 40% fat spreads made using these oils are roughly the same. This confirms the earlier statement the melting characteristics of the lowfat spreads are not determined by the melting characteristics of the oils used in their formulation.

### Effect of the amount of fat

The second major consumer trend mentioned above is toward ultralow fat levels. As the amount of fat is reduced in a water-in-oil emulsion, the emulsion is expected to be increasingly unstable. At about 74% by volume of aqueous phase the water droplets are almost touching each other (12). This occurs at about 28% fat by weight, assuming the densities of the aqueous and oil phases differ by about 10%.

In addition to the stability problems that increase as the amount of fat in a spread is lowered, there also are "meltability" problems. The lowfat spreads do not seem to melt as readily in the mouth, thus giving an unpleasant coating. In addition, the flavor release characteristics are adversely affected. Table 2 shows the Mettler drop points of margarines and spreads of different fat levels. It is seen that the 80% fat margarine "melts" near body temperature. Even at 50% fat, the drop point is near body temperature. As you go down to 40% fat, the drop point rises sharply. The rise is even more rapid at still lower fat levels. Note that the hardness of the oil blend in these products was the same.

## CONCLUSION

We have shown in three separate ways - the absence of an effect of oil phase crystallinity and a strong effect of the amount of the oil phase and of the drop size on the drop point of spreads - that the Mettler drop point is a rheological concept. The Mettler drop point correlates with meltability of spreads in the mouth, which governs the flavor release characteristics in these products. Emulsion rheology is known to be a strong function of dispersed phase volume fraction and of the droplet size. As the dispersed phase volume fraction approaches the close-packed limit the viscosity and elasticity are known to increase rapidly (5). If the droplets were not deformable and were all of uniform size, the emulsion would "freeze" and act like a solid. Since the water droplets are deformable and the emulsion is polydispersed, the spread behaves as a plastic solid at the close-packed limit.

## ACKNOWLEDGEMENT

We would like to acknowledge Kraft General Foods for permission to publish this manuscript and also to our coworkers, particularly K.J. Sternberg, S. Huntoon, A.G. Gaonkar and F.J. Sasevich, for their help in carrying out this investigation.

## LITERATURE CITED

(1) Andersen, A.J.C. and P.N. Williams, Margarine, Pergamon Press, London (1965).

(2) Haumann, B.F., New Ideas for Margarine, INFORM, 1(3) 174-182 (1990).

(3) Walker, R.C. and W.A. Bosin, Comparison of SFI, DSC and NMR Methods for Determining Solid-Liquid Ratios in Fats, JAOCS, 48(2) 50-53 (1971).

(4) Gaonkar, A.G., Interfacial Tensions of Vegetable Oil/Water Systems: Effect of Oil Purification, JAOCS, 66(8) 1090-1092 (1989).

(5) Sherman, P., Rheological Properties of Emulsions, in Encyclopedia of Emulsion Technology, P. Becher, Ed., pp 405-437, Marcel Dekker, New York (1983).

(6) Hinze, J.O., Fundamentals of the Hydrodynamic Mechanism of Splitting in Dispersion Processes, AIChE J., 1(3) 289-295 (1955).

(7) Shinnar, R., On the Behavior of Liquid Dispersions in Mixing Vessels, J. Fluid Mechanics, 10 259-275 (1961).

(8) Daily Food and Drink Report, L. Gasparello, Ed., 2(65), King Communications Group, Washington, D.C. (1990).

(9) INFOSCAN, Information Resources Inc., Publisher, Chicago. (1990).

(10) Haumann, B.F., New Ideas for Margarine, INFORM, 1(3) 182-184 (1990).

(11) Vold, R.D., Mittal, K.L., and A.U. Hahn, Ultracentrifugal Stability of Emulsions, in "Surface and Colloid Science, Volume 10", E. Matijevic, Ed., Plenum Press, New York (1978).

(12) Princen, H.M., Rheology of Foams and Highly Concentrated Emulsions, J. Colloid and Interface Sci., 91(1) 160-175 (1983).

Table 1. The effect of thickener on the average water droplet sizes and the Mettler Drop points of 40% fat spreads

| Thickener? | Average Water Droplet Size | Drop Point |
|---|---|---|
| Xanthan Gum | 13.9μm | 51.5°C |
| ---- | 7.1μm | 68.9°C |

Table 2. Mettler drop points of margarines and spreads.

| % Fat | Soft/Stick | Drop Point, °C |
|---|---|---|
| 80 | Stick | 34.1 |
| 80 | Soft | 34.0 |
| 50 | Soft | 37.0 |
| 40 | Soft | 46.8 |

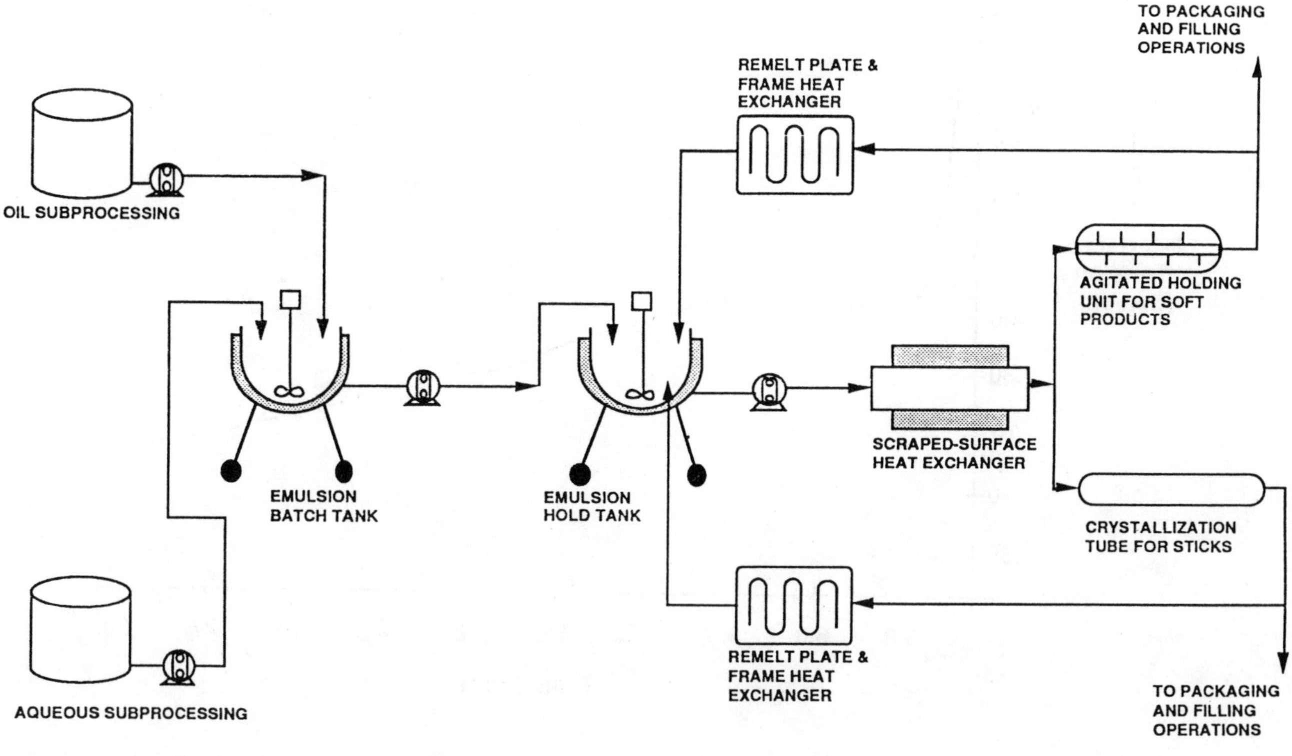


Figure 1. Schematic of a typical manufacturing process for margarines and spreads.

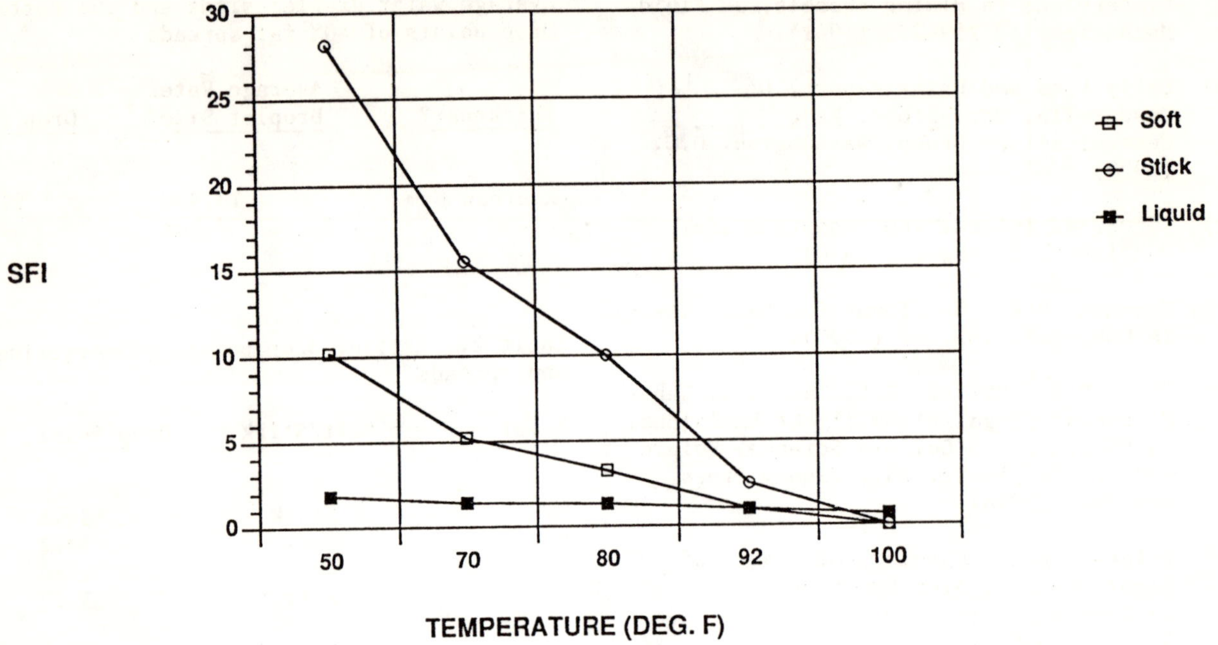


Figure 2. Solid fat index (SFI) of oil blends as a function of temperature.

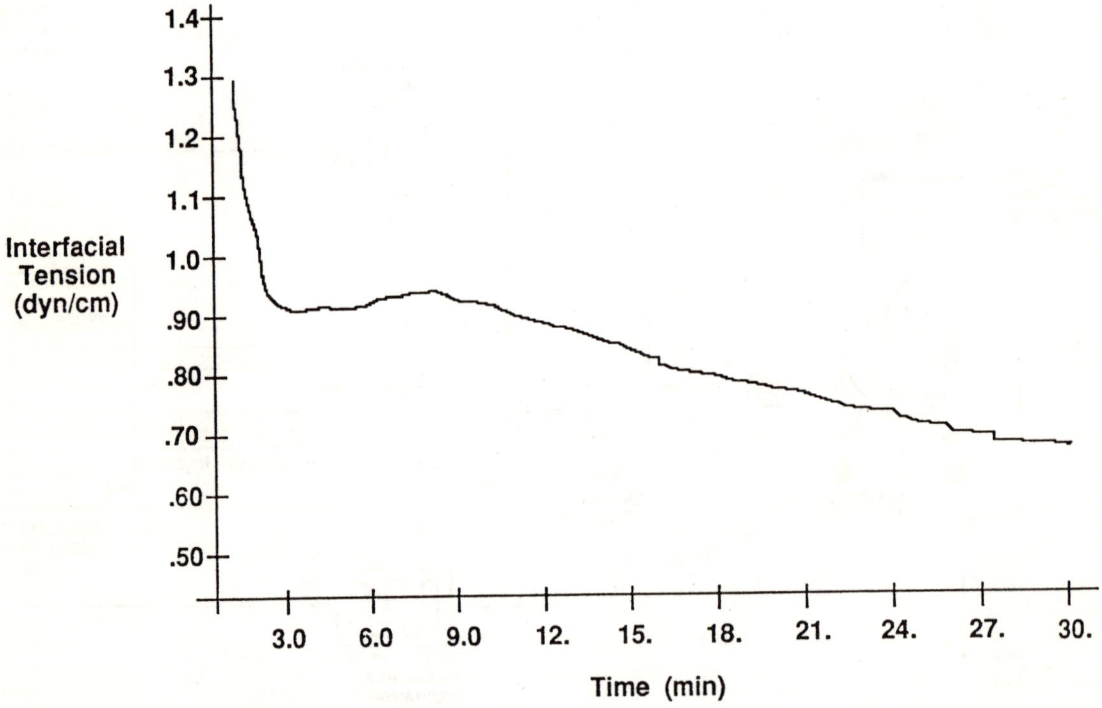


Figure 3. Interfacial tension @ 50 °C of oil-water system containing 0.51% lecithin in the oil phase and 3.6% NaCl in the aqueous phase.

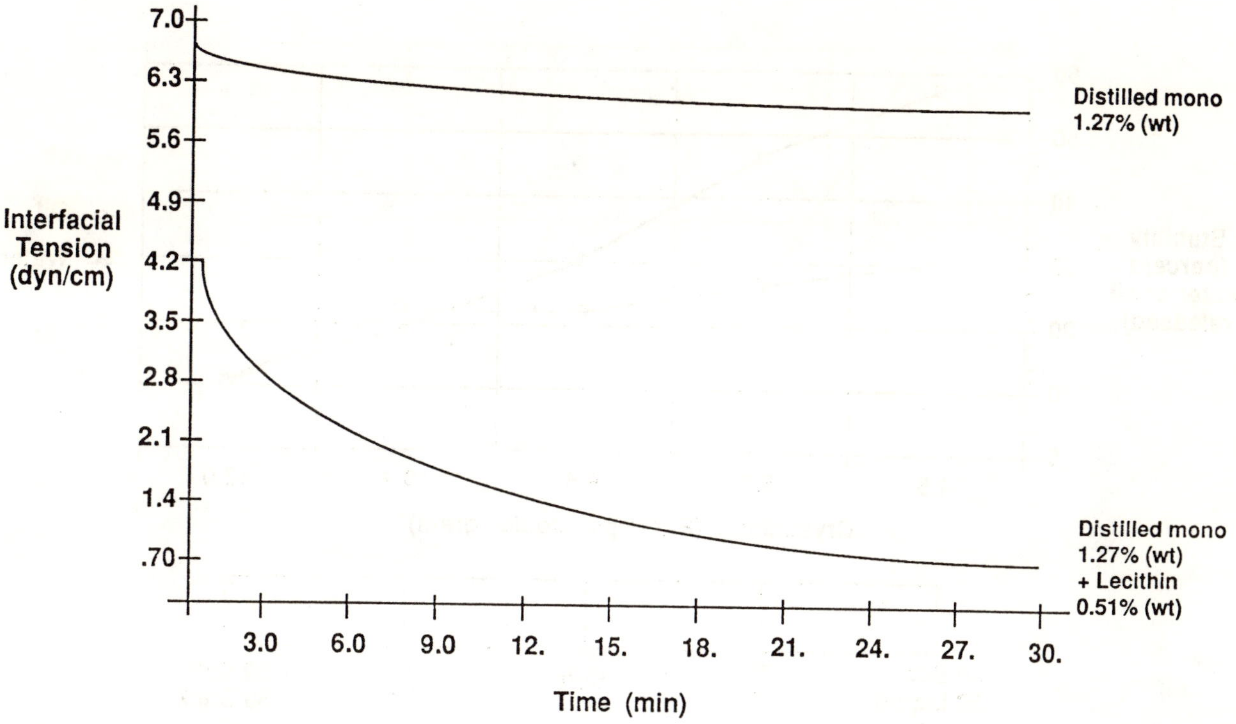


Figure 4. Interfacial tension @ 50 °C of oil-water systems containing (a) 1.27% distilled mono (b) 1.27% distilled monoglyceride + 0.51% lecithin in the oil phase and 3.6% NaCl in the aqueous phase.

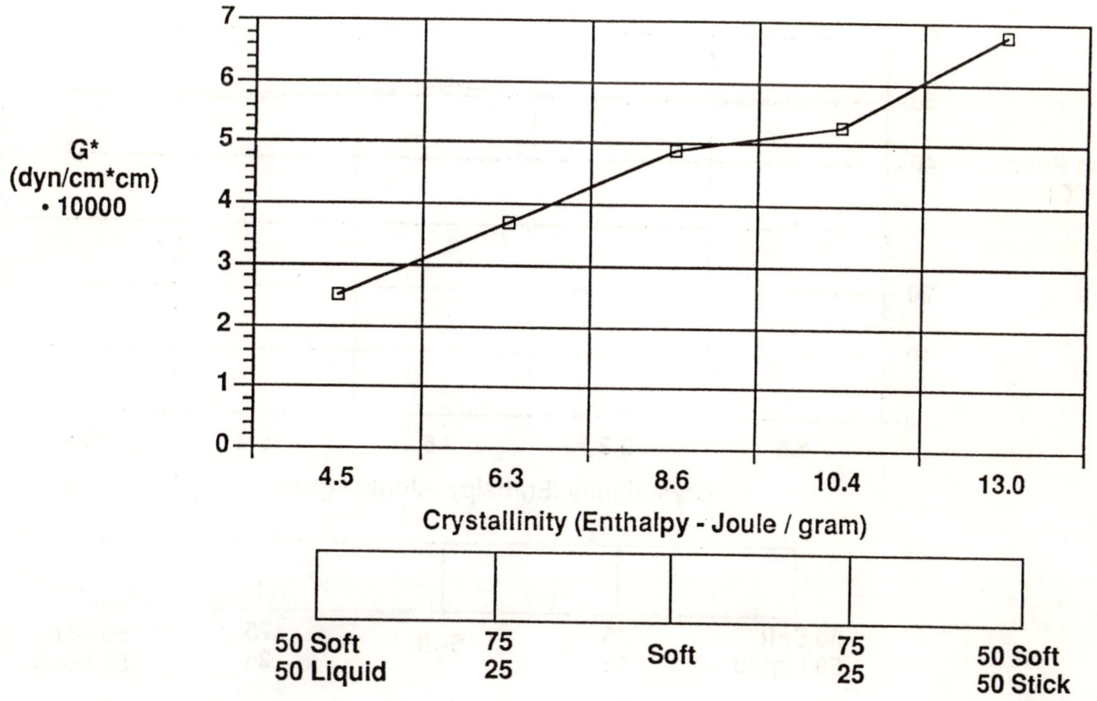


Figure 5. Effect of the degree of crystallinity on the firmness of 40% fat spread as measured by G* from oscillatory measurements.

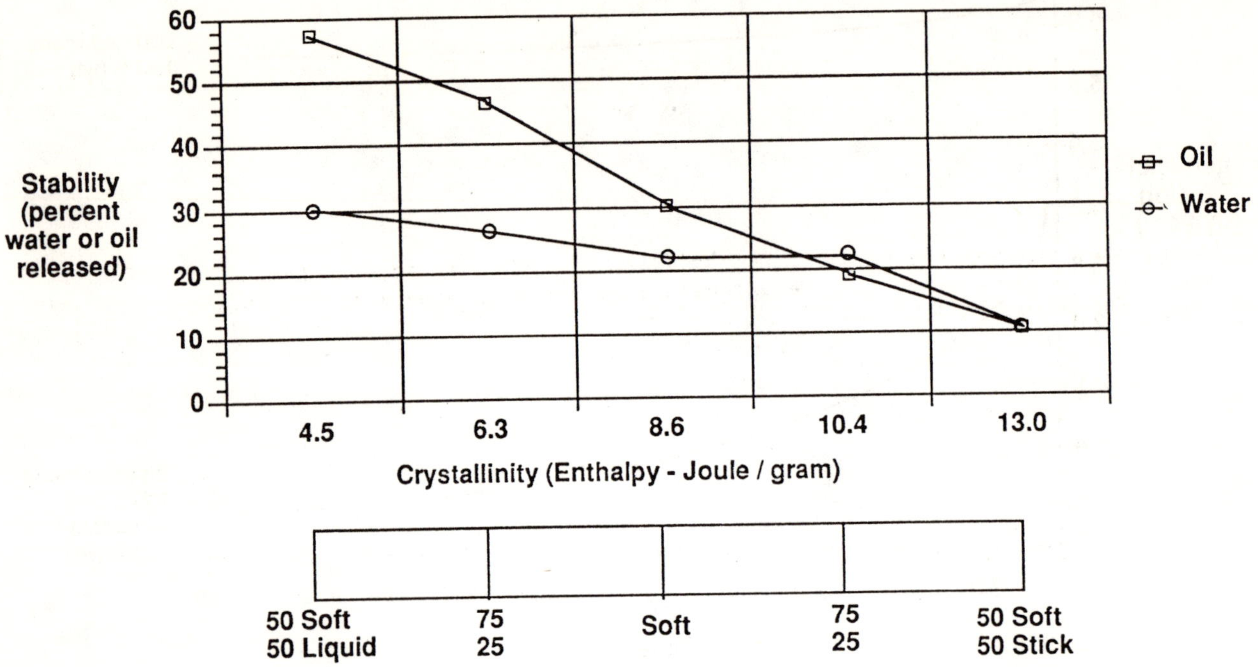


Figure 6. Effect of the degree of crystallinity on the emulsion stability of 40% fat spreads as measured by ultracentrifugation.

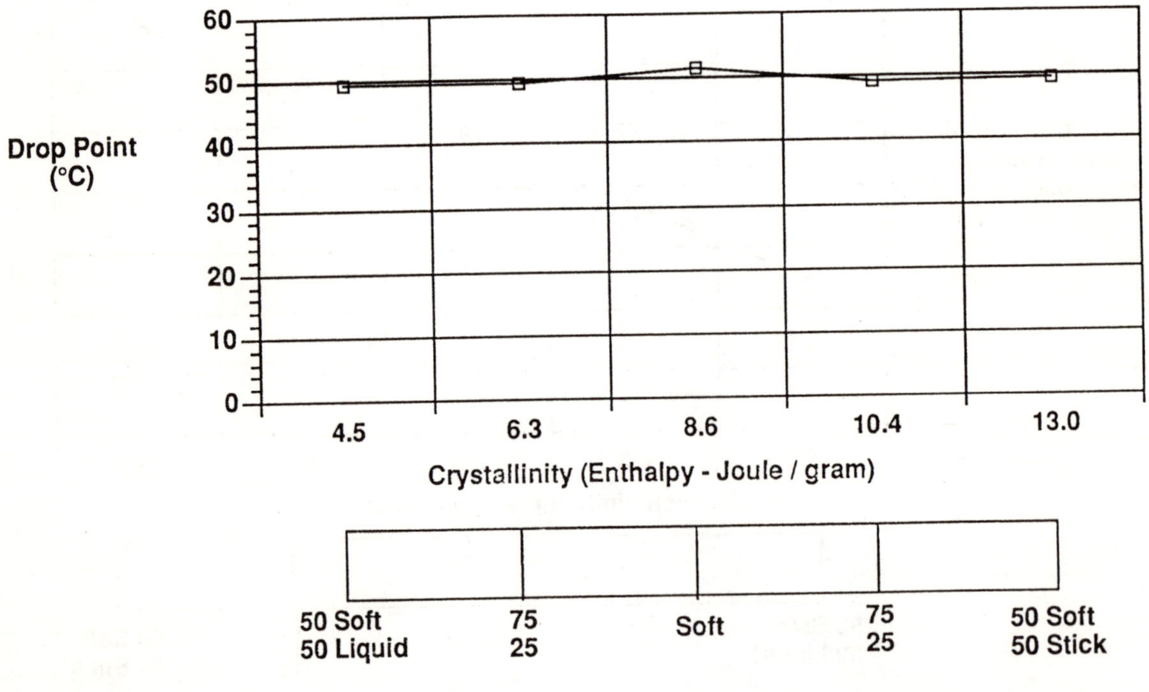


Figure 7. Effect of the degree of crystallinity on the Mettler drop point of 40% fat spreads.

# CHARACTERISTICS OF ICE CREAM

David S. Reid ■ Department of Food Science, University of California, Davis, Davis, CA 95616

Ice cream is a complex, multiphase system. It is in part an air containing foam, and so an understanding of foam properties is important. The continuous phase of the foam is, of itself, an emulsion, requiring therefore an appreciation for emulsion science. As a further level of complexity, the effect of temperature induces liquid-solid phase change, which affects system properties markedly, e.g. by concentrating the unfrozen part of the aqueous phase. Due to the system being disperse, nucleation of the solid phase must be considered an important factor, too. The paper attempts to outline the important factors which must be considered in order to better understand the manufacture and properties of ice cream.

The purpose of this paper is to raise questions, rather than to answer questions. The purpose is to illustrate the complexity of ice cream, and to indicate the variety of descriptions, and the variety of models, which must be taken into account when trying to explain the characteristics of ice cream. The paper is not intended to be a comprehensive review, rather it is hoped that it will serve as a catalyst toward future studies of an intriguing problem.

There is, at the present time, a resurgence of interest into the properties of food emulsions, and many of the papers presented in this symposium have been accounts of the state of the art. In a related symposium, the properties of food foams are under discussion. Ice cream could appropriately be discussed at either symposium, since ice cream is both a foam and an emulsion. It will therefore be appropriate to consider, during this presentation, the additional complexities which might be added to emulsion science by considering a foam, and the additional complexities which might be added to foam science by considering an emulsion.

As indicated, ice cream is both a foam and an emulsion. Fairly complete descriptions exist of the properties of model foams and of model emulsions, so what is to prevent us from simply applying these to the ice cream system? Apart from the added complexity of mixing a foam and an emulsion, what other complicating factor needs to be considered? The answer to this is temperature. The properties of the ice cream mix will be very temperature dependent, and this will add a whole new dimension of complexity to the problem of providing a comprehensive scientific description of the properties of an ice cream system. We will therefore discuss aspects of the ice cream system in the absence of the complicating effect of temperature, and then look at the role of temperature in the properties of the real ice cream system.

## THE ICE CREAM FOAM

Ice cream is a foam. What is a foam? For our purposes, a foam can be described as a dispersion of air in a continuous phase. Some models exist which make an attempt to explain the properties of foams. These apply primarily to very simple systems. In the ice cream foam, the continuous phase is primarily a sugar solution, but in this solution there are other components which have a profound effect upon the behavior of the system. For example, there are proteins present which may be surface active, and affect the stability of the air - aqueous phase interface. This will affect the stability of the air bubbles, and also the initial size of the air bubbles. Problems which have to be addressed include the rate of drainage of the foam, the rate of diffusion of air through the continuous phase etc. The amount of protein and other surface active agents will be very important. Most practical ice cream formulations also include polysaccharide stabilizers in the aqueous phase. These macromolecules, as one of their functional characteristics, will influence foam stability, and foam drainage, since they alter the viscosity of the aqueous phase. The overall composition of the aqueous phase will be very important to the characteristics of the foam. We will consider the properties of the aqueous continuous phase further when we consider the effect of temperature on the ice cream foam system.

## THE ICE CREAM EMULSION

Ice cream is an emulsion. An emulsion can be described as a dispersion of one phase in another. Models and theories exist which explain the behavior of many simple emulsions. The properties of both continuous phase and dispersed phase are very important. Surface effects at the interface also play a major role in emulsion formation, and in emulsion stability. For ice cream the dispersed phase is an oil/lipid phase, and the continuous phase is aqueous. The fact that this aqueous phase is itself a foam will provide complications. The stability of this emulsion will depend upon many factors, including the properties of the lipids, and the presence of small molecule and large molecule surface active materials in the aqueous phase. The possibility that air-lipid interfaces may exist will play a role in stability. Many of these factors have been described in the papers discussing the fundamentals of food emulsions. I wish, here, to consider primarily the role of temperature, as it is this which makes the problem of describing the ice cream system so much more difficult. Temperature will affect the properties and stability of the foam, and also the properties and stability of the emulsion.

## THE ROLE OF TEMPERATURE

Consider the three main constituent phases of ice cream mix, viz air, aqueous phase, and lipid phase. Each of these phases will behave in a characteristic manner as temperature is lowered. An understanding of the effect of temperature on the properties of the ice cream system requires that we consider the effect of temperature on each of the constituent phases, and on the interactions which take place between the phases.

### The air phase

As temperature is lowered, the air phase will reduce in volume, according to the gas laws. This will alter the phase volume of air in the foam, thus changing the rheological properties of the foam, and also reducing the amount of air-aqueous interface. Thus the properties of the interface will be affected, and the composition of the interfacial layer may change. The number of surface active molecules at the interface may reduce, or if there were insufficient surface active molecules to give full coverage at the higher temperature, coverage will increase as the amount of interface decreases. The temperature at which the air is incorporated into the foam will clearly be very important.

### The aqueous phase

The aqueous phase will become more

viscous as temperature is reduced, and constituent molecules, in particular proteins, may undergo conformational change as a result of the lower temperature. This could affect their surface active characteristics. The increased viscosity could play an important role in the ease of formation of a foam, and in the air bubble size in the foam. It is well documented in model systems that the ease of formation, and stability of a foam, are affected by continuous phase viscosity.

As the temperature continues to reduce, ice crystallization becomes possible. The ice nucleation capabilities of the aqueous phase, and the ability of the phase to undercool, will affect the final outcome. As ice crystallizes out, the remaining aqueous phase between ice crystals becomes more viscous, due to increasing concentration. This affects the ease of formation of a new foam. Also, the increasing concentration of this phase affects the properties of its constituent molecules, both in bulk, and at the air-solution interface. The conditions under which ice will form, and the ice crystal size, and concentration profiles during processing will all be complex consequences of the system used for heat removal. The temperature of the scraped surface heat exchanger of the ice cream machine, together with the rate of mixing at this surface, and rate of dispersal of material into the bulk mix, will largely control the nucleation, and subsequent propagation of ice. As the ice grows, the sugar concentration will increase, and polymer concentrations will increase. The viscosity increase mentioned will be due both to the increased concentrations, and to the lower temperature. The presence of solid ice particles will also have an influence on the stability of the foam, as it is well known that the presence of particulate material in the continuous phase may alter the stability, and drainage characteristics of a foam.

Because of these complex, temperature linked changes, the time of introduction of air into the mix to make the foam can be very important. The surface properties of the proteins will change, due to both the reduced temperature, and also to the increased ionic strength of the solution phase. Diffusional properties will change, so the characteristics of the foam will be very different, due to the altered properties of the continuous phase. In some circumstances other solutes might precipitate from the aqueous phase, influencing foam stability. So far we have considered the aqueous phase as the continuous phase of the foam. It is, of course, also the continuous phase of the emulsion. the increased concentration, increased viscosity, and existence of ice, will all affect emulsion properties. before we consider this further, however, we must examine the effect of temperature on the disperse phase of the emulsion.

<u>The lipid phase</u>

As with the other phases, the lipid phase, too, is markedly affected by temperature. Two distinct situations must be considered. In bulk, the lipids used in ice cream have a complicated solidification pattern, with a considerable percentage becoming solid at temperatures close to, but above, the freezing point of water. Various crystal forms exist. These are well documented for simple lipids, as is the temperature and time dependence of their interconversion under different processing histories. For dairy fats, the crystallization picture is not yet so well described. Dairy fats exhibit a complex time and temperature dependent crystallization behavior. This can influence the properties of an emulsion prepared by dispersing this lipid material in an aqueous continuous phase. The time temperature history of the bulk fat will be important. the temperature of emulsification will also be important.

As complex as the situation is with

bulk fat, a new level of complexity enters when we consider emulsified fat. The patterns of nucleation of crystallization in a dispersed phase differ markedly from those in a bulk phase. Propagative crystal growth is confined to individual droplets. The crystal forms after a given temperature time history can differ markedly from those in a bulk preparation. The properties of the droplets will be critically dependent upon the percentage of solid fat in the droplet, and on the crystal forms of this fat.

In a normal emulsion, the dispersed liquid phase is in the form of approximately spherical droplets. the flow behavior depends on droplet size, and phase volume. As the fat crystallizes, the spheres will change shape. this will affect flow properties. If shear is applied during the crystallization step, the resulting emulsion can have very different properties. Consider the additional complication of having ice crystals in the continuous phase. these crystals may affect the shear fields experienced by the dispersed fat. The stability of the dispersed phase may depend upon the extent of crystallization, and the degree of distortion of the individual "droplets." Some of the fat, rather than be part of this dispersed phase, may end up as a surfactant molecule at the air - solution interface.

## CONCLUSION

The coexistence of foam and emulsion, together with the presence of ice, makes for a problem of great complexity. Simple models may lead to insights about different aspects of the behavior, but in the end the complexity of ice cream is such that studies have to be performed on the complete system. The complexities of fat nucleation and crystallization in bulk or dispersed phase, ice nucleation and growth in a stirred system, and air dispersion as a foam in such a complex medium, will make it difficult for experiments on ice cream to be replicated at a variety of centers, so the studies of models will be necessary in conjunction with the real system. A second level of more complex models needs to be investigated. The simple models which suffice for probing emulsions, or probing foams, are too far removed from the real world complexity of ice cream, though they do lead to insights which can be amplified upon in a set of second generation models.

# EMULSION FILMS ON FOOD PRODUCTS TO CONTROL MASS TRANSFER

John M. Krochta ■ Department of Food Science and Technology, University of California, Davis, CA 95616

Multicomponent edible films, especially those containing a lipid as moisture barrier and a polar high polymer as cohesive structural matrix, have much potential for maintaining food quality. Several emulsion film systems have been and are presently being investigated, and a commercial emulsion film system is available. Emulsion films provide unique opportunities for controlling mass transfer, including control of moisture movement, oxygen transfer and preservative migration. The microvoid model attributes mass transfer of gases and vapors through emulsion films to the formation of fine continuous microvoids, which are formed between the microparticles of hydrophobic material and the matrix of high polymer as the emulsion dries to form a coating membrane. The micropathway model attributes mass transfer through emulsion films to the high polymer matrix iteslf.

Foods quickly lose fresh character as their moisture content evaporates to the air, or when moisture moves from one component of the food to another, or as oxygen diffuses into the food to enhance respiration or oxidation. Certain foods also lose quality when they experience aroma, flavor or color loss, either to the atmosphere or to other food components. Still other foods suffer quality loss when they absorb too much fat or oil during frying or cooking. The transfer can occur from the cooking fat or oil, or from other food ingredients. Each of these events involves undesirable mass transfer, and leads to deterioration of food texture, flavor, color, aroma or nutritional value. The results include reduction of shelf life and food value, and reduced ability of the food industry to meet increasing consumer demands for higher-quality food products.

Thin, transparent and nearly invisible edible films from natural ingredients formed on food surfaces or between food components could help prevent such product quality loss. Edible films coating the surfaces of foods or separating components of more complex food products can prevent quality deterioration by acting as barriers to control product moisture transfer, oxygen uptake, volatiles and solutes loss, or lipid transfer.

In addition to controlling mass transfer, edible films can act as carriers of food ingredients and additives. These can include various flavors, colors, aromas, nutrients, antioxidants and antimicrobials. In some cases edible films might become quite visible and discernible, and thus be an aid to the creation of new food products. In the case of antioxidants and antimicrobials, edible films can concentrate these additives at the surfaces of foods and prevent them from migrating into the food. In this manner edible films again control mass transfer and maintain the levels of preservatives required for effective action with a minimum amount of these materials added.

Finally, edible films can help improve food handling. They can do this by providing a smooth surface so that foods can slide over each other and slide into packages more easily. Edible films can also increase the structural integrity of foods which tend to be fragile and disintegrate easily.

Thus, a proposed definition for an edible film is the following: "A thin continuous layer of edible material placed on or between food components to provide a barrier to mass transfer (e.g., moisture, oxygen, volatiles, solutes, lipids) and/or provide a carrier of food ingredients and additives, and/or improve food handling." An example of a potentially useful edible film would be a coating on apple slices which reduces moisture loss, reduces respiration (and increases shelf life) by controlling transfer of oxygen and carbon dioxide into and out of the apple slices, carries antibrowning and antimicrobial agents, and provides a smooth surface which allows easy handling and packaging.

## EMULSION FILMS

The concept of edible films is not a new one. Two recent reviews describe concepts, applications and research ranging over a considerable period of time (1, 2). Relatively new are attempts to utilize some new and improved materials, and to formulate multicomponent systems for edible films which attempt to maximize the advantages and minimize the disadvantages of each component.

### Emulsion Film Systems

Both Guilbert (1) and Kester and Fennema (2) pointed out the potential of multicomponent systems for effective edible films, especially those containing a lipid as a moisture barrier and a polar high polymer as a cohesive, structural matrix. Such systems can help overcome the application (adherence and homogeniety), mechanical (breaks and pinholes) and organoleptic (waxy taste and brittle character) problems of pure lipid films (1). Schultz et al. (3) were perhaps the first to study a film of this type, combining lipids with a low-methoxyl pectinate film. They found that lipid-pectinate films were better moisture barriers when the lipids were added as surface coatings to pre-formed pectinate films compared to being suspended in the original film-forming pectinate aqueous solution. Waxes were found to be the most effective lipid materials studied. Beeswax and paraffin wax coated onto glycerol-plastized pectinate films reduced moisture transmission by 90% and 99%, respectively, compared to uncoated films. When beeswax and paraffin wax were suspended in the film forming solution they reduced moisture transmission by 80 and 95%, respectively, compared to plain glycerol-plasticized pectinate films. Pectinate films coated with vegetable fat, lauric acid or bile acids had moisture transmission rates 75-90% lower than uncoated films. Suspending lauric acid in a pectinate film reduced moisture transmission by 65%, but suspending vegetable fat and bile acids had little effect. Interestingly, crosslinking pectinate films with a solution of calcium chloride which contained emulsified lipid generally produced films with lower moisture transmission than uncrosslinked pectinate films containing suspended lipid.

Ukai et al. (4) developed a concept which involves films prepared from emulsions or suspensions of hydrophobic liquids or solids in aqueous solutions of water-soluble high polymers. Emulsifiers such as lecithin, sucrose fatty acid esters and alkali salts of oleic acid were used. Upon drying, films containing microparticles of the hydrophobic materials in matrixes of the high polymers were formed. Hydrophobic materials examined included coconut oil, hydrogenated coconut oil, soybean oil, beeswax and carnauba wax. Water-soluble high polymers investigated included casein, gelatin, starch, sodium carboxymethylcellulose, acacia (gum arabic) and sodium alginate. Coating fruits with these emulsions gave less variation with time in product weight, brix, and citric or malic acid content and slowed ripening, compared to uncoated fruits. Coating vegetables produced less product weight loss and quality loss.

Kamper and Fennema (5-7) developed edible films including lipids in two different ways: 1) coating various lipids onto hydroxypropyl methylcellulose:polyethyleneglycol (HPMC:PEG) films (coating technique), and 2) forming films from systems which included various lipids dissolved in ethanol-water solutions of HPMC with PEG (emulsion technique). As the ethanol-water solvent evaporated in the latter technique, an emulsion-type system likely developed. The resulting films were bilayers, with a lipid layer which had separated from the emulsion resting on a HPMC layer. The coated films displayed a trend of decreased permeability with decreased fluidity of the lipid material used. Compared to a HPMC:PEG film, moisture transmission rates of coated films were reduced by amounts ranging from 29% with a corn oil layer to over 99% with beeswax and paraffin wax. Using the emulsion technique, a bilayer film with a lipid layer of a stearic acid-palmitic acid was created with a 99% reduction in moisture transmission rate compared to a HPMC:PEG film. This compared to a 90% reduction when a larger amount of the stearic acid-palmitic acid blend was coated onto the HPMC:PEG film. In addition, the bilayer film formed using the emulsion technique was more flexible and resistant to mechanical damage than the coated film. More recent work by Kester and Fennema (8, 9) has focused on composite films prepared by coating beeswax onto films comprised of methylcellulose (MC), HPMC and the stearic acid-palmitic acid blend. Additional work by Greener and Fennema (10, 11) has involved development of composite films prepared by coating beeswax onto films of MC. Wax coating has allowed these films to function at high relative humidities.

Vojdani and Torres (12) have studied the ability of various lipid-containing films to concentrate the antimicrobial potassium sorbate at the surfaces of foods. They used film formulations and film-forming techniques similar to those described earlier (5-11). These films functioned to maintain the concentration of potassium sorbate necessary for antimicrobial activity at the food surface by retarding its diffusion into the food bulk. Permeability constant determinations indicated that surface resistance to microbial growth could be enhanced significantly.

Guilbert (1) investigated films whose formation involved emulsification of an ethanolic solution of stearic acid, palmitic acid and carnauba wax in aqueous solutions of casein or gelatin. The resulting oil-in-water emulsions were placed on surfaces, solutions of tannic or lactic acid were added, and the resulting mixtures were dried to form films. These films apparently resembled the bilayer films produced by the emulsion technique used by Kamper and Fennema (5-7), giving a lipid layer and a protein layer. The resulting films were described as having good water barrier properties, but being brittle, not transparent, and having a waxy taste.

Krochta et al. (13) investigated development of emulsion films involving addition of various lipids to casein films. Stable emulsions were formed from aqueous solutions of sodium caseinate with acetylated monoglycerides, beeswax, carnauba wax or paraffin wax, and were used to produce films upon drying. Treatment of the resulting emulsion films with calcium ascorbate buffer adjusted to the isoelectic point of the casein rendered the films insoluble and decreased their permeability to moisture. Buffer-treated emulsion films with equal amounts of acetylated monoglyceride and casein had moisture permeabilities one-fourth that of plain casein films. Doubling the amount of acetylated monoglyceride relative to casein reduced moisture permeability by another 50%.

The commercial edible coatings Pro-long and Semperfresh have been the most studied multi-component film systems which incorporate lipids and a water-soluble high polymer. Pro-long and Semperfresh both contain sucrose polyesters, carboxymethyl cellulose and mixed mono-and di-glycerides. This mixture is dispersed and then applied as an aqueous suspension by dipping, spraying or coating, followed by drying. These coatings have been shown to control water loss and respiration, and to affect ripening and quality factors, in several different whole fruits (14-23). Actual permeabilities of these films to moisture, oxygen, carbon dioxide and ethylene have not been determined for comparison to other film systems.

## Emulsion Film Models

Multicomponent systems such as emulsion films provide the opportunity to tailor-make films to the needs of the food being coated. In the example described earlier, the apple pieces must be protected from moisture loss and must have respiration slowed in order to provide a practical shelf life. Edible films have the potential for reducing oxygen transfer to, and carbon dioxide transfer from, such fruit and vegetable products to slow respiration. However, to continue the earlier example, various apple varieties have different optimum oxygen and carbon dioxide levels for optimum shelf life extension. Other fruit and vegetable products also have different optimum levels for shelf life extension. Thus, the need for adjusting film properties to allow for different oxygen and carbon dioxide mass transfer clearly exists.

Microvoid Model. The emulsion films developed by Ukai et al. (4) involved dispersions of waxes and other hydrophobic materials in aqueous solutions of water-soluble high polymers, such as proteins and polysaccharides. The emulsion film compositions and process defined in their work involved a) a ratio of hydrophobic material to water-soluble high polymer substance in the range of 10:1 to 10:5; and b) the hypothesized development of fine continuous microvoids, formed between the microparticles of hydrophobic material and the matrix of water-soluble high polymer as the composition dried to form a coating membrane. Ukai et al. (4) never verified the existence of such microvoids in the dried emulsion film. However, they believed that it was through these microvoids that the minimum amount of oxygen required for respiration action to avoid physiological damage was supplied and that discharge of carbon dioxide and moisture was controlled.

If the Microvoid Model is correct, it would seem that the extent of such microvoids could be controlled by a) altering the composition of the emulsion film, and/or b) modifying the size of the microparticles inbedded in the high polymer matrix. The former would involve the relative amounts of the hydrophobic and high polymer materials. The latter would involve the degree of emulsification and the nature of any added emulsifiers. Different relative amounts of hydrophobic material and high polymer and different degrees of emulsification will provide different areas of interface for microvoid formation. Such manipulation could allow tailor-making a particular emulsion film to match the oxygen and carbon dioxide transfer needs of a particular fruit or vegetable product.

Micropathway Model. An alternative to the Microvoid Model which may describe the functioning of some and maybe all emulsion films is the Micropathway Model. In the Micropathway Model, the emulsion films also involve dispersion of hydrophobic materials in aqueous solutions or dispersions of high polymers. When the film composition dries, the emulsion film contains microparticles of the hydrophobic material suspended in highly concentrated polymer material. The high polymer material provides a matrix which consists of narrow, continuous micropathways around the hydrophobic microparticles. The nature of the films described by this model is such that microvoids between the microparticles of the hydrophobic material and the high polymer matrix pathways do not occur when the film is dried. Thus, transmissions of oxygen, carbon dioxide and/or moisture through such a film on a food product do not occur through microvoids, but through the high polymer matrix itself. This transmission through the high polymer matrix occurs because high polymers like proteins and polysaccharides are often quite compatible with moisture and gases and can offer little resistance to their transmission. However, the narrow, micropathways which result from the drying of the film provide tortuous transmission of gases and moisture around the large number of hydrophobic microparticles.

If the Micropathway Model is correct, a considerable number of opportunities exist for controlling respiration, or oxidation, and/or moisture loss in foods. The extent and size of micropathways could be controlled by a) altering the composition of the emulsion film, and/or b) modifying the size of the microparticles inbedded in the high polymer matrix. The former would again involve the relative amounts of the hydrophobic and high polymer materials. The latter would again involve the degree of emulsification and the nature of any added emulsifiers. Such manipulation could again allow tailor-making a particular emulsion film to match the oxygen and carbon dioxide transfer needs of a particular fruit or vegetable product.

The number and diameter of high polymer pathways can control transmission of gases and moisture in emulsion films, as described above. The length of high polymer pathways, as determined by emulsion film thickness, can also function to control transmission. In addition, the particular properties of the high polymer used in the emulsion film composition will influence gas and moisture transmission. Each high polymer has unique transmission properties and each high polymer can be modified in specific ways to modify these properties. For example, the molecular configurations of proteins can be modified by simply adjusting pH or divalent ion content of emulsion systems. More complex protein modifications involve chemical or enzymatic grafting or crosslinking. Thus, the unique aspect of the Micropathway Model, should it prove true, is the opportunity to control the gas and/or moisture transmission of a film by 1) choosing different high polymers for making the emulsion film, or 2) modifying the high polymer material selected with chemical and/or enzymatic action. Krochta et al. (13) found that adjusting the pH of casein-acetylated monoglyceride emulsion films to the isoelectric point improved the moisture barrier properties of these films. This apparently occurs because, in adjusting a protein to its isoelectric point, the number of positive and negative charges on the protein are balanced and the protein can assume a different molecular configuration. In some cases, the change in molecular configuration is such that the formerly water-soluble high polymer protein is rendered insoluble. These changes in molecular structure can apparently modify the protein's ability to provide resistance to gas and moisture transmission in an emulsion film.

## SUMMARY AND CONCLUSIONS

Edible films can provide an opportunity to control mass transfer in foods, as well as a carrier of food ingredients and an aid to food handling. Multicomponent films have the advantage of allowing use of the functional properties of several components.

If and when an emulsion film conforms to the Microvoid Model, the respiration (or oxidation), and/or moisture loss of the food upon which it is placed can be controlled by selecting and controlling the following emulsion film properties: 1) the ratio of the hydrophobic material to the high polymer, 2) the size of the hydrophobic microparticles in the emulsion film, and 3) the thickness of the film.

If the emulsion film conforms to the Micropathway Model, the particular oxygen, carbon dioxide and moisture transmission properties of the high polymer material constituting the matrix passageways are also critical to the performance of the film. The transmission properties of the high polymer material pathways are established by the particular high polymer chosen for making the film composition, or by alteration of the high polymer being utilized by chemical and/or enzymatic modification.

Considerable work must be done to: 1) determine where the Microvoid and Micropathway Models are appropriate for emulsion-type films, and 2) exploit the opportunities for the use of emulsion films in maintaining and improving food quality.

LITERATURE CITED

1. Guilbert, S., "Technology and application of edible protective films," in Food Packaging and Preservation, Mathlouthi, M. (Ed.), Elsevier Applied Science Publishing Co.,London (1986).

2. Kester, J. J. and O. R. Fennema, Food Technol., 40 (12), 47 (1986).

3. Schultz, T. H., J. C. Miers, H. S. Owens and W. D. Maclay, J. Phys. Colloid Chem., 53 (9), 1320 (1949).

4. Ukai, N., S. Ishibashi, T. Tsutsumi and K. Marakami, U. S. Patent No. 3,997,674 (Dec. 14, 1976).

5. Kamper, S. L. and O. Fennema, J. Food Sci., 49 (6), 1478 (1984).

6. Kamper, S. L. and O. Fennema, J. Food Sci., 49 (6), 1482 (1984).

7. Kamper, S. L. and O. Fennema, J. Food Sci., 50 (?), 382 (1985).

8. Kester, J. J. and O. Fennema, J. Food Sci., 54 (6), 1383 (1989).

9. Kester, J. J. and O. Fennema, J. Food Sci., 54 (6), 1390 (1989).

10. Greener, I. K. and O. Fennema, J. Food Sci., 54 (6), 1393 (1989).

11. Greener, I. K. and O. Fennema, J. Food Sci., 54 (6), 1400 (1989).

12. Torres, J. A., J. Food Proc. Pres, 13, 417-430 (1989).

13. Krochta, J. M., A. E. Pavlath and N. Goodman, "Edible films from casein-lipid emulsions for lightly-processed fruits and vegetables," in Proceedings of the Fifth International Congress on Engineering and Food, Spiess, W. E. L. and Schubert, H. (Eds),Elsevier Science Publishers Ltd, London (1990).

14. Banks, N. H., J. Expt. Bot., 35 (150), 127 (1984).

15. Smith, S. M. and J. R. Stow, Ann. Appl. Biol., 104, 383 (1984).

16. Banks, N. H., Acta Hort., 157, 105 (1984).

17. Banks, N. H., Scientia Hort., 24, 279 (1984).

18. Chu, C. L., HortScience, 21 (2), 267 (1986).

19. Smith, S, J. Geeson and J. Stow, HortScience, 22 (5), 772 (1987).

20. Drake, S. R., J. K. Fellman and J. W. Nelson, J. Food Sci., 52 (5), 1283 (1987).

21. Meheriuk, M. and O. L. Lau, J. Amer. Soc. Hort. Sci., 113 (2), 222 (1988).

22. Motlagh, F. H. and P. C. Quantick, Intl J. Food Sci. Tech., 23, 99 (1988).

23. Dhalla, R. and S. W. Hanson, Intl J. Food Sci. Tech., 23, 107 (1988).

# PROPERTIES OF STABILIZING COMPONENTS IN FOAMS

J.B. German ■ Department of Food Science and Technology, University of California, Davis, Davis, CA

Foams are thermodynamically unstable colloidal systems in which gas is maintained as a distinct dispersed phase in a liquid matrix. Both drainage of the much denser liquid phase and bubble coalescence are otherwise spontaneous events that must be kinetically hindered for foam stability. Drainage between lamella and through the plateau borders is impeded by fluid viscosity and interaction with the film surfaces The kinetic barrier to bubble coalescence and rupture is provided by the dynamic interfacial viscosity and elasticity of the surfactant or polymeric films surrounding the gas bubble. The major goal in understanding the stability of foams is to reconcile the molecular properties of components in the fluid and interfacial films with their macroscopic effects on foam stability. While considerable research has been conducted on the molecular basis of foaming properties of small surfactants, these are not readily translatable to higher density foams stabilized by large polymers such as proteins. The consensus of recent research suggests that the ability of polymers to form a coherent gel-like highly elastic film at the bubble interface is critical for foam stability. Consequently, the capacity of certain components to disrupt the integrity of existing films becomes important in the destabilization of these same foams. Methodologies to examine the molecular and interfacial properties of foam components in parallel with macroscopic measurements of foam stability and drainage are essential to further advances in this field.

## INTRODUCTION

Foams are representative of the physical state of many foods and similar biomaterials(1). Bread, cakes, ice cream, etc are at some point in their development characterized by a profusion of air cells in a continuous aqueous phase (2) (3). In a more precise physical description they are thermodynamically unstable colloidal systems in which gas is maintained as a distinct dispersed phase in a continuous liquid matrix. There is considerable energetic disadvantage in the large surface area of air water interface leading to the thermodynamic instability of foams (4). Foams are thus metastable in that they are prevented from lowering their net energy by a kinetic i.e. activation barrier provided by the surfactant material at the interface. The presence of surfactant is essential to the formation and immediate stability of bubbles. Pure water does not foam. In some cases foams are the desired target and the properties of density and structure essential to the final product. In other cases foams represent an unwanted consequence of agitation which can even preclude certain processing methods. This paper will review the principles of foam formation and describe certain properties of the surfactant materials present in foamable foods that provide the stability of this intriguing microscopic and macroscopic state.

Department of Food Science and Technology, University of California, Davis

In addition to forming films some components including some proteins interfere with and even disrupt the integrity of existing films. Short chain alcohols are well characterized and applied as foam breakers. The structural basis of these net disruptive interactions is not well known but in practice this is critical in many food applications. It is also possible that a greater understanding and prediction of these events could lead to many useful applications of reversing the foaming process.

## Foam Stability, Macroscopic Events

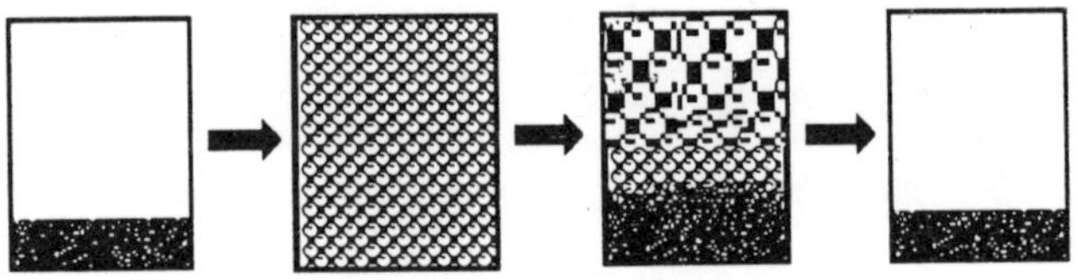

The macroscopic progress of foam formation and collapse is schematized above. A gas is introduced into a continuous liquid, forming a relatively homogeneous gas dispersion. Air is incorporated into a liquid by one of three processes; whipping in which the

liquid is actively forced around external gas, sparging in which gas is actively forced through liquid, and in situ generation in which gas is generated within the liquid itself. In each case the formation of a bubble with a stabilized gas/liquid interface or film is required. Otherwise, bubbles collapse into each other instantly. Even when foam is successfully formed, the processes of liquid drainage/bubble creaming and bubble collapse lead to increasing vertical stratification of the foam, eventually leading to a return to the liquid state.

There are thus two steps that are necessary for most practically stable food foams. There must be sufficient surfactant available to form a foam in the first place. For most foods however, this is not enough. Food foams are almost invariably 'Kugelschaum', that is holding a high liquid phase content. Most fundamental studies address the properties of 'Polyederschaum' foams, those with very low water content and polyhedral bubbles. Kugelschaum foams are macroscopically unstable due to the large difference in density between the continuous and discontinuous phases. Thus analagous to emulsion particles bubbles of diameter d will cream, or rise spontaneously according to stokes law.

$$\text{Velocity} = \frac{g\,\Delta\rho\,d^2}{18\,\eta_0}$$

The large density difference $\Delta\rho$ between the continuous phase and the air bubbles leads to rapid bubble velocities and the sole means to prevent this process is to increase the effective viscosity $\eta_0$ of the continuous phase. For reasonable stability the continuous phase must exhibit a yield value (5). In practice, food foams are gelled or immobilized by network formation in the continuous phase. The bubbles themselves must be stable until components in the continuous phase are induced to gel. This increase in continuous phase viscosity or yield value affords long term stability. The critical question therefore is film stability. If a surfactant component can maintain discrete bubbles long enough for continuous phase agents to develop stabilizing interactions, the component is a usable foaming agent. If the bubbles break initially, or collapse in response to the conditions necessary to induce gelation of the continuous phase (i.e. heating), the component is not competent for foaming in that application. Therefore, the second step necessary for stability of the bubbles in most food foams is the maintenence of bubble structure through this immobilization stage.

One of the challenges to foam study has been the inability to distinguish liquid drainage from bubble collapse. Although these two events typically proceed simultaneously this is not necessarily true. Ideal foaming components will permit drainage without extensive bubble collapse. Methods which measure foam stability by the loss of liquid from the bottom of the foam are unable to distinguish these two events. We have developed non-invasive NMR imaging techniques (6), (7) to monitor stability in real time throughout the foam. Two foams illustrating quite different stabilities are shown below. Beer foam is an example of a low density foam in which a relatively unstable bubble structure is necessary. Both liquid drainage and bubbble collapse occur simultaneously. This behavior is illustrated by the vertical hydrogen signal intensity profiles from top to bottom of the foam as a function of time. This technique essentially follows the drainage of liquid as it progresses from the foam above into the liquid beer below. An important observation is that the foam density rapidly drops to zero at the top of the foam as it drains indicating that the entire bubble structure of the foam has collapsed.

**Drainage of Beer Foam by NMR Image Analysis**

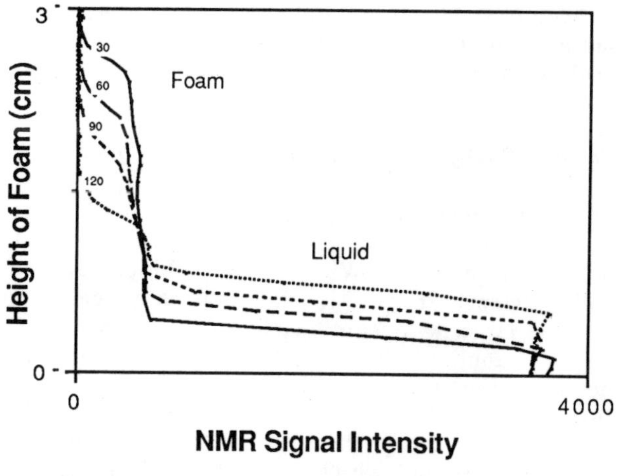


**NMR image analysis of beer foam collapse. 30 ml aliquats of commercial beer were sonicated to generate an approximately equal volume of foam. The sample was placed in the imaging coil of a GE 2T imaging spectrometer and vertical signal projections recorded at 30 second intervals according to German and McCarthy 1989.**

This is contrasted with the profiles from egg white foam. At the top of the foam as drainage proceeds the intensity of hydrogen signal decreases. But it does not drop to zero. The bubble structure is intact and a significant hydrogen signal is retained in the gradually decreasing density of the foam. These results clearly reflect the quite disparate macroscopic collapse of these prototypical foams.

**Collapse of Egg White Foam**

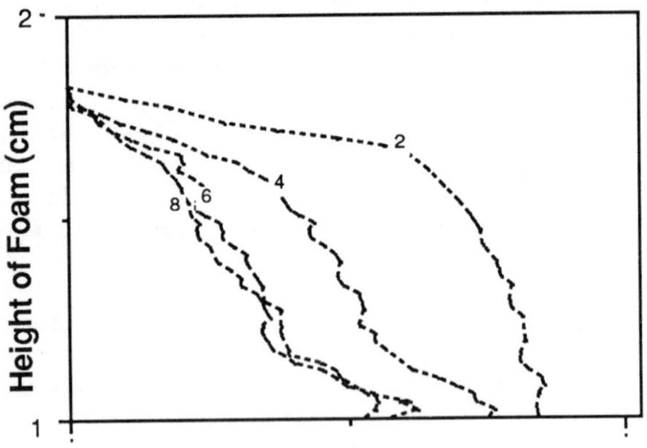


**NMR imaging of egg white foam. Egg white from commercial eggs was whipped for 5 minutes in a beater style mixer and a 10 * 4 cm cylinder sample punched from the foam. Vertical projections were recorded as described above.**

## Foam Stability, Microscopic Events

To evaluate foam stability it is necessary to understand the actual processes that cause bubble rupture. When two bubbles approach and surfaces begin to interact they can repel each other and remain stable or collapse into one larger bubble. It is generally agreed that this occurs through hole nucleation. Once a hole has formed it will grow rapidly and spontaneously due to decreasing the net interface. Therefore, any mechanism which would tend to prevent close approach of films or the formation of a nucleating hole would stabilize the film. Conversely, components which accelerated the nucleating process either by providing nucleating sites (bridging) or diluting repulsive components would destabilize the film. In foods the kinetic barrier to bubble coalescence and rupture is typically provided by a film surrounding the bubbles. The complexity of molecular structure of proteins permits them to act both as the initial surfactant forming the foam and as the polymer matrix stabilizing them. The variation in performance of different proteins reflects the varied interactions necessary to participate in these events and the non-uniformity of protein structure at both the sequence and three dimensional levels. Other components however also act either to form and stabilize foams or to destabilize them including polysaccharides, simple sugars, lipids and charged or uncharged polymers. These will be discussed below.

The development of intermolecular associations at the interface leads to alterations in surface properties that have measurable rheological consequences. That is, the surface takes on gel like attributes and these can be described by two dimensional variables which are direct analogies to three dimensional rheological terms. The most relevant for film and bubble stability is the dynamic dilatational modulus. This is measured as the change in surface tension as a function of change in surface area.

**Dynamic Dilatational Modulus** $$\varepsilon = -\frac{\delta \gamma}{\delta \ln A}$$

Clearly a surfactant film which is very dilute or contains few sites of intermolecular interaction will display little effect of changing surface area, while a cohesive film with many interactive sites will exhibit significant elasticity. In a great variety of model systems a high dynamic or elastic modulus correlates with increased stability (8), (9), (10)

To understand foam stability it is necessary to determine the interactions between surface components which lead to a stable film. Thus we need to know how the forces enjoined on soluble components during the dynamic interfacial exposure and absorption period alter their structure, and then how these altered structures lead to novel molecular and macromolecular interactions. It is the novel associations which then provide the film properties and ultimately an energetic barrier to resist desorption and collapse. The consensus of recent observational research is that the ability to form a gel-like film at the bubble interface is indeed critical for food foam stability (11)(12)(13).

## FOAMING COMPONENTS IN FOODS

Since simple surfactants such as soaps are effective in the formation of foams, the requirements are primarily for an amphiphilic molecule which can diffuse to the interface, lower the surface tension, accumulate to a surface concentration that provides an effective film and resist the recollapse of adjacent air bubbles. Stability of food foams however requires maintenence of bubble films. Different components exhibit wide variations in this property.

### Proteins

It is the resolution of a certain paradox that provides proteins with their unique capability in stabilizing surface films. That is, proteins are soluble in water because they don't associate yet they function to stabilize films because they do. How can this be? In solution proteins as polymers are typically globular and highly folded and are prevented from interacting largely by a charged and hydrophilic surface providing a net repulsion through electrostatic and dehydration energies. The maintenence of this structure is critical since unfolded glubular proteins typically aggregate readily. In foam formation air is mechanically driven into the aqueous solution and surface is generated with a time scale on the order of 10's to 1000's of square cm per ml per second. In response to this appearance of large surface areas, soluble proteins are subjected to an interfacial exposure/adsorption. For most proteins which owe their specific solution conformation to solvent interactions (14), this appearance of surface provides an opportunity to lower their net free energy by altering their structure (12). While the proteins may not associate with each other in the soluble state this unfolding or alteration in structure allows for novel associations in the interface (15).

The film formation process can be divided into four steps. The generation of new interface via the introduction of air is the necessary, high energy, initial event. This can be accomplished by whipping air into solution, sparging an air stream into a liquid or generating gas directly within the liquid itself. While all three generate air bubbles they differ in the rate of surface expression and in the agitation of the interface once formed. In step two, soluble components, typically proteins, maintained in a particular conformation primarily by entropically favorable solvent (hydrophobic) interactions arrive at this interface by diffusion. They must then diffuse into the interface itself. While it was historically thought that arrival at the interface was sufficient for surface adsorption this has recently been shown to be a naive assumption. The probability of surface adsorption is not necessarily one. The various solvation energies favoring the solution phase may impose an activation energy to the adsorption process itself. Srinivasan Damodaran (16) has provided both a theoretical treatment of the process as well as documentation (17) of rate behavior of surface adsorption of different proteins.

Conceptually, proteins adsorb to available interface with a probability related to the energetics of the surface and to their structure and conformation in solution. According to this treatment, there is a potential energy function associated with an amphiphilic component in the subphase which dictates the probability of surface adsorption. This can be readily seen by its effect on the diffusion coefficient of macromolecules where changing interfacial tension is the measured parameter. For certain proteins the lack of adsorption for some collisions leads to an erroneously low value for the diffusion coefficient (18).

Once absorbed, the elimination of solvent energetics from the balance of its conformational stability drives the protein to unfold and 'pursue' more energetically favorable conformations. As a consequence, proteins will form novel associations first intramolecularly (step three), via changes in conformation. This leads to a phenomenon termed surface denaturation wherin the conformations adopted in the interface are not reversible by resolvation (12). The molecular flexibility of different proteins is apparently different in that highly stabilized proteins do not unfold to a great extent at the interface and do not surface denature. These proteins are also generally poor foam formers (19)(20).

As more protein arrives and unfolds at the interface, associations between proteins become increasingly probable. These intermolecular interactions (step four) are critical for subsequent stabilization of the suface films of bubbles. Forces and structures which favor inter-molecular associations in this step improve foaming ability while forces and structures which prevent inter-molecular associations in this step decrease foaming ability (13), (3).

The importance of the retention of some native protein conformation and structure in the development of foams is well documented by both model systems and actual food foams(13), (11). Perturbation of native structure significantly alters the interfacial behavior of proteins (21). Foaming is perhaps the most sensitive functional property to alterations in the structure of dairy proteins (22). The ability to associate in a denaturing environment, i.e., an interface, and yet retain some residual conformation appears to be crucial. Extensive modification, including glycosylation, succinylation and enzymatic hydrolysis reduce secondary structure. While these treatments have frequently increased surface adsorption kinetics and surface pressure, film strength and foam stability were almost invariably decreased (13) (3), (11).

Those proteins which most owe their solution structure to solvent entropy, would unfold most readily after leaving the solvent and adsorbing to a surface. The formation of novel intermolecular associations would be more favored and desorption from the surface less likely. In contrast, those proteins which are stabilized in solution predominantly by intramolecular interactions (ion pairing, disulfide bonds, etc) even though they could spontaneously adsorb to a clean interface, would not as readily alter their native structure and continue to form novel intermolecular associations. These proteins thus have a higher desorption tendency. For example, soy 11 s globulin is stabilized in solution by both hydrophobic associations and intramolecular disulfide bonds and is relatively resistant to heat and urea denaturation (23). Cleavage of the disulfide linkages does not in itself dissociate the protein but significantly reduces the stability to dissociation in response to a solvent destructuring denaturant. The reduction of these intramolecular disulfide bonds dramatically enhances both film strength and foam stability (9). Alternatively, foam stability of solutions of beta lactoglobulin which readily unfolds at an interface is not affected by the reduction of disulfide bonds (24).

The forces which favor or retard the interaction of proteins unfolding at an interface are the most critical to foam stability but are the least understood. This is a technical problem since the stability of protein films presumably requires multiple layer protein associations beyond that which basic studies of monolayers can readily investigate. Although hydrogen bonding and hydrophobic interactions likely play a role here, electrostatic interactions have been most studied. Proximity to the isoelectric pH is the most widely observed environmental correlate to foaming ability, in contrast to emulsion stability which is maximized away from the isoelectric point (11)(25)(9)(26). Such studies have been widely interpreted to indicate that a strong repulsive force developed as a consequence of adsorption of highly charged proteins to dilute monolayer coverage prevents the approach of additional, like-charged proteins (27)(28). Furthermore, one would predict that masking these charges with counterions would reduce electrostatic repulsion and favor protein associations beyond a monolayer (27). Salts as neutralizing counterions are indeed highly effective protein foam enhancers at pH's distal from the isoelectric point (see 1) but are actually depressants at the isoelectric point (25)(9). Thus at the isoelectric point, the forces of repulsion are minimized and the forces of attraction can be observed to contribute to film formation.

A balance of electrostatic interactions are therefore thought to be critical for foaming. Too much and repulsive forces preclude the formation of protein films. Too little and film strength is compromised. Again, soy glycinin has proven a useful model protein. The native structure consists of acidic and basic subunit fragments which interact in the native state through hydrophobic interactions and disulfide bonds. If these disulfide bonds are cleaved and the structure specific hydrophobic associations disrupted, e.g. by heat or urea, the oppositely charged subunits will spontaneously reassociate, but now via electrostatic interactions (23)(29). The protein fragments on surface absorption would thus exhibit the ability to associate via electrostatic interactions in the developing interfacial film only if the intramolecular disulfide bonds were cleaved. This was found to be true. The sensitivity to electrostatic repulsive forces was demonstrated by the fact that salt as a counterion was still required for the disulfide reduced protein to form stable films far from the isoelectric point of the protein.

The ability of basic or positively charged proteins to act as foam stabilizers is perhaps the most exciting development of this understanding of the forces involved in film integrity (30). The basic protein lysozyme has long been considered to be important to the whipping properties of egg white (31). Hen egg lysozyme was added at 0.5% to solutions of 5 % beta lactoglobulin, an acidic protein already exhibiting good foam stability. This addition had no significant effect on the surface pressure of the protein solution, however, the foam stability (time to 50 % drainage) was more than doubled (24). Similarly positively charged proteins have been shown to increase stability of foams to the presence of lipids (30) and to heating (Phillips unpublished).

## Carbohydrates

Carbohydrates and particularly polysaccharides are widely used for ingredient functionality in disperse food materials. They have been most studied in their role as viscous and gelling components. As strongly hydrated molecules, carbohydrates tend to alter the rheological properties of water. Thus both sugars at high concentration and polysaccharides at low concentration increase the viscosity of the continuous aqueous phase in foams. This has an immediate and obvious effect on increasing the stability of foams by reducing the drainage rate as described above. Polysaccharides have also been observed to contribute to foam formation directly. Carbohydrates exhibit similar macromolecular properties to proteins in possessing both hydrophilic and hydrophobic regions on the same molecule. This causes the polymer to orient itself at a new interface as do all amphiphiles to minimize the total free energy of the system. This surfactant property allows foaming. Detailed studies have not been carried out to determine the ability of pure polysaccharides to form stable films at the surface of bubbles. This may be possible through the relatively wide capacity for modifying the polymer chain. It has been observed that both the number and distribution of modifications to polysaccarides influence interfacial tension and foaming behavior. Gums and celluloses adsorb to an interface and lower the surface tension thus promote foam formation in highly purified model systems . Thus, interfacial tension decreased with increasing hydroxypropylation of cellulose (32) and foam stability of guar gum foams increased with increasing hydroxyalkylation (33).

The area of most intense research in the role of polysaccharides in foams is in their role as modifications to proteins. It has been widely observed that glycoproteins have considerable interfacial film stability (34), (35), (36). Similarly the addition of carbohydrate to beta lactoglobulin progressively increased the surface film viscosity and stability of foams generated from protein solutions (37).

In view of the dramatic improvements in film and foam stability provided by oppositely charged proteins described above it is suggested that a possible role for ionic polysaccharides in stabilizing films of oppositely charged proteins appears promising. The role of polysaccharides in specifically complexing and stabilizing to a bubble film may be substantial. In such mixtures the ionic properties of the polysaccharide could be tailored to the specific protein composition. This basic mechanism has been identified in the stabilization of beer foam by alginates (38). Carboxyl groups on the alginate were found to interact with amino groups on the surface bound proteins forming stabilizing cross-links. The further possibility that these stabilizing interactions could be activated by the same processes favoring gelation in the continuous phase is appealing.

## Fats and Emulsion particles

The role of fats and emulsion particles in foams actually spans the entire spectrum from foam stabilization to depression. Foam retardants or breakers function by intercalating into the interfacial film network and preventing or disrupting existing associations (5). Films provide stability to foams by virtue of a continuous physical and kinetic barrier to coalescence. As surfactants, amphiphilic lipids have been recognized as potent foam depressants. Simple lipids serve to compete effectively with proteins and other polymeric surfactants for available surface. Because they tend not to form stable films this effective dilution of the film forming components will reduce surface viscosity and dilatational moduli and thus act passively to reduce stability of bubble films (2).

The presence of lipids as destabilizing components is not always apparent. Lipids may act as foam depressants when complexed to proteins in solution. A foam depressant activity associated with a high molecular weight protein was found in variable quantities in acid and sweet wheys (24). It is likely a lipoprotein derived from the milk fat globule membrane. Polar lipids which are the actual foam depressant at the level of the interface, are transported and released as a function of the structure of the lipoprotein as well as that of the protein film onto which it adsorbs (39).

Emulsion particles have been well studied for their foaming effects in foods largely due to the relevance to dairy systems (40). These particles also add a significant level of complexity to the problem however. Fats and oils which are insoluble in aqueous systems exist in foods and beverages as dispersed emulsion particles surrounded by a stabilizing emulsifier. Consequently, fats contribute to three physical states: 1 the physical particles themselves which alter the continuous phase viscosity, the bubble lamellar shape etc, 2 an amphiphilic (frequently charged) interface consisting of a large total surface area capable itself of interacting with bubble surfaces, proteins, of contributing ions, buffering capacity, etc. and 3 a nonpolar medium of variable fluid/solid fat composition providing three dimensional structure, preferentially solubilizing other nonpolar compounds, interacting directly with an air water interface etc.

Macroscopic particles have been observed to both stabilize and destabilize foams depending on the foam and the properties of the particle. The stabilization of foams by protein aggregates has been interpreted to be that of solid particles (41). Even carbon particles have been observed to stabilize foams to collapse. In each case the mechanism of action has been proposed to be repulsion between approaching bubble film surfaces. These interpretations are valid for primarily hydrophilic or charged particles. Emulsion particles are generally themselves stable because of a charged or highly polar surface, therefore they are presumably able to stabilize foam lamellae again by preventing the close approach of bubbles (40). In contrast, hydrophobic particles in general and crystalline fat suspensions in particular have significant and direct antifoaming activity. The mechanism of this action is thought to be through a bridging mechanism between lamellae in which nucleation of a connecting hole is the outcome (42). The size and composition of emulsion particles appears to be critical as well, since large, partially crystalline milk fat globules which are incompressible, destabilize foam lamellae purportedly due to physical destruction of the lamella (43), (44).

Surfactant materials from emulsion surfaces depress foam stability for the reasons described above (45). Phospholipids and partial glycerides effectively displace proteins from bubble surfaces and thus reduce film integrity. Both whipping cream and ice cream are less stable when made from milks which have been abused during processing to increase lipolysis.

The role of the emulsion triglyceride core in foam stability remains poorly understood in spite of extensive investigations. This is partly due to the presence of liquid and solid fat and the complexity of crystallization behaviours of many functional food emulsions including dairy emulsions. The best example of fat's potential role in foam stability is through the bridging of adjacent bubbles with solid fat which confers the long term stability of whippped cream (46). Thus, liquid fat leaking from the globule during whipping (43) readily crystallizes into a solid three dimensional network (46) providing the yield stress necessary for long term stability of the foam (5). Thus both solid and liquid fat in the core of the particle may be necessary for this effect. The crystals promote globule rupture (44) and the liquid fat promotes clumping of unruptured globules and gas bubbles (45). It is also possible that some of the fat globules contain super-cooled liquid fat which rapidly crystallizes on breaking the emusion (47).

## METHODOLOGY

The many variables which have an impact on foaming emphasize the necessity of adopting standardized or at least comparable methodologies for foam development and measurement. No single laboratory can hope to resolve the various interacting components into a predictive model of functionality. If methodologies are quantitatively comparable important observations can be incorporated into all models for foam stability.

At the practical level certain descriptive analyses are still lacking. Non-invasive measurements of structure development and collapse are essential. In this regard novel imaging techniques will prove particularly useful. Similarly, it is absolutely necessary to extend studies of foam stability to include the response to gelling processes such as heating. The inability to maintain foam structure through this phase of food foam processing would absolutely preclude its functionality regardless of its activity in model systems. Results discussed above argue that certain unifying principles may dictate stability of bubble films during heating.

The most obvious descriptive property of foams which needs to be accurately measured is their bubble size distribution. Without this fundamental parameter, mechanistic studies are difficult to interpret. Development of automated analyses for these measurements will advance the field significantly.

Most theoretical studies and mathematical models on foam formation, stability and rheology have assumed foams of very low density. While these are useful for certain conceptual and mechanistic problems they fail to describe the properties of foams relevant to foods in which the density is invariably high. Although preliminary attempts may be largely empirical, any quantitative predictions which can be tested against actual food systems will be valuable. Studies to provide clues to the type and magnitude of forces acting between surface components especially at the interface itself are critical.

Acknowledgement: Support, in part by the California Dairy Foods Research Center.

LITERATURE CITED


1. Aubert, J. H., Kraynik, A. M. and Rand, P. B. Scientific American 74-83. (1986)

2. Bikerman, J. J. Foams. Springer Verlag, New York. (1973)

3. Kinsella, J. E. and Whitehead, D. M. in 'Proteins at Interfaces' ACS symposium series, pp. 343, Amer. Chem. Soc., Wash DC. (1987)

4. de Feijter, J. A. "Thermodynamics of Thin Films". in 'Thin Liquid Films' Surfactant Science Series v 29, (Ivanov, Eds.), pp. 1, Marcel Dekker, New York. (1986)

5. Prins, A. in 'Food Emulsions and Foams', (E., D., Eds.), Royal Society of Chemistry, Burlington House, London, UK. (1987)

6. German, J. B. and McCarthy, M. J. J. Ag. Food Chem. 37, 1321-1324. (1989)

7. Heil, J., McCarthy, K., German, J. B., McCarthy, M. J. and Patino, H. J. Brew. Sci. in press, (1990)

8. Huang, D. D., Nikolov, A. and Wasan, D. T. Langmuir 2, 672-677. (1986)

9. German, J. B., O'Neill, T. E. and Kinsella, J. E. J. Am. Oil Chem. Soc. 62, 1358-1368. (1985)

10. Djabbarah, N. F. and Wasan, D. T. AIChE Journal 31, 1041-1043. (1985)

11. Halling, P. J. CRC Critical Review, Food Sci. Nutr. 155, 13. (1981)

12. MacRitchie, F. Adv. Colloid Interface Sci. 25, 341-385. (1986)

13. Phillips, M. C. Food Technol. 35, 50-60. (1981)

14. Tanford, D. The Hydrophobic Effect. Wiley Interscience,, New York. (1973)

15. Norde, W. Adv. Colloid Interface Sci. 25, 267-340. (1986)

16. Damodaran, S. "Film Properties of Modified Proteins". in Food Proteins, (Soucie, W. and Kinsella, J.E., Eds.), pp. 88, Am Oil Chem Soc, Champaign Il. (1989)

17. Damodaran, S. and Song, K. B. Biochim Biophys Acta 954, 253-64. (1988)

18. Ward, A. and Tordai, L. J. Chem. Phys. 14, 453. (1946)

19. Graham, D. E. and Phillips, M. C. J. Colloid Interface Sci. 70, 403. (1979)

20. Graham, D. E. and Phillips, M. C. J. Colloid Interface Sci. 70, 427. (1979)

21. Song, K. B. and Damodaran, S. J. Agric. Food Chem. 35, 236. (1987)

22. Liao, S. Y. and M.E., M. J. Food Sci 52, 1033. (1987)

23. German, J. B., Damodaran, S. and Kinsella, J. E. J. Agric. Food Chem 30, 807-812. (1981)

24. Phillips, L. "A Study of the Foaming Properties of Proteins", Ph.D. Thesis,Cornell University Ithaca NY., (1988)

25. Kim, S. H. Ph.D. Thesis,Cornell University Ithaca NY, (1984)

26. Kim, S. H. and J.E., K. J. Food Sci. 52, 128. (1987)

27. Norde, W. Adv Colloid Interface Sci 25, 267-340. (1986)

28. Kinsella, J. E. Food Chem. 7, 273. (1981)

29. Damodaran, S. and Kinsella, J. E. J.Agric. Food Chem. 30, 1249. (1981)

30. Poole, S., West, S. and Walters, C. J. Sci. Food Agric. 35, 701. (1984)

31. Aschaffenburg, R. and Drewery, J. Biochem. 65, 237. (1957)

32. Sarkar, N. Polymer 25, 481-489. (1983)

33. Prabhanjan, H., Gharia, M. M. and Srivastava, H. C. Carbohydrate Polymers 12, 1-7. (1990)

34. Schmidt, R. D. Adv. Biochem Eng. 12, 41. (1979)

35. Sharon, N. and Lis, H. Chem. Eng. News March 30, 21-25. (1989)

36. Canton, M. C. and Mulvihill, D. M. in Proc Int. Dairy Fed. Symp., pp. 339-345, (1983)

37. Waniska, R. D. and Kinsella, J. E. "Glycosylation of b lactoglobulin and surface active properties". in Food Proteins, (Soucie, W. and Kinsella, J.E., Eds.), pp. 100-131, Am Oil Chem Soc., Champaign, Il. (1989)

38. Jackson, G., Roberts, R. T. and Wainwright, T. J. Inst. Brew. 86, 34-37. (1983)

39. German, J. B. and Phillips, L. "Molecular Properties of Proteins important for foams". in Food Proteins, (Soucie, W. and Kinsella, J.E., Eds.), pp. 132-143, Am. Oil Chem. Soc., Champaign, Il. (1989)

40. Walstra, P. and R., J. Dairy Chemistry and Physics. John Wiley, New York. (1984)

41. Hermansson, A. M., Sivik, B. and Skjoldebrand, C. Lebensmittel Wiss. Technol. 4, 210. (1971)

42. Aronson, M. P. Langmuir 2, 653-658. (1986)

43. Mulder, H. and Walstra, P. The Milk Fat Globule. Pudoc, Wengeningen, The Netherlands. (1974)

44. van Boekel, M. and Walstra, P. Colloids Surf. 3, 109. (1981)

45. Anderson, M. and Brooker, B. E. "Dairy Foams". in Advance in Food Emulsions and Foams, (Dickinson, E. and Stainsby, G., Eds.), Elsevier, the Netherlands. (1988)

46. Precht, D. "Fat Crystal structure in cream and butter". in 'Thin Liquid Films' Surfactant Science Series v 29, (Ivanov, Eds.), pp. 305, Marcel Dekker, New York. (1986)

47. Simoneau, C. and German, J. B. Unpublished, (1990)

# INTERNAL LIQUID FLOW DURING FOAM DRAINAGE: COMPARISON OF THEORY AND EXPERIMENT

Kathryn L. McCarthy and Julianna R. Heil ■ Departments of Food Science and Technology and Agricultural Engineering, University of California, Davis, Davis, CA 95616

Magnetic resonance imaging (MRI) was used to measure the vertical mass distribution in an aqueous foam system as a function of position and time. A preliminary macroscopic drainage study identified a time frame to evaluate liquid drainage in the foam system. The signal intensity of hydrogen nuclei was recorded by MRI for a protein stabilized foam (reconstituted egg white). This intensity directly correlates to water density in the foam. Foam density measurements, recorded over a time frame of twenty minutes, showed distinct stratification during liquid drainage. The drainage profiles were compared to a theoretical model for transient drainage of a persistent foam. Experimental data show much greater stratification than predicted by this model.

An aqueous foam is a dispersion of gas in a continuous liquid phase with liquid as a minor component by volume. These foams are important in a wide variety of uses, such as drilling fluids in the petroleum industry, fire fighting agents, effluent processing, froth flotation, and applications in the food and pharmaceutical industries (1, 2, 3). The stability of foam is a critical parameter and in the past, has been frequently characterized by a drainage rate. A foam is formed and then left alone in a vessel. Gradually some liquid collects at the bottom of the container; its volume is measured after several time intervals and the volume exuded, V, is plotted against time, t. Typically, this type of drainage curve has been modeled by first order rate kinetics (4). However, the information obtained from this type of study does not elucidate the physics of foam drainage.

Magnetic resonance imaging (MRI) provides a non-invasive method to measure foam density and drainage rates as a function of both time and position without disturbing the structure of the foam (3, 5). MRI data reported by German and McCarthy (1989) presented qualitative signal intensities as a function of time and position. Preliminary comparison of MRI density profiles and theoretical predictions was done for a water/olefin surfactant system by Assink, et al. (5).

In this study, the contribution of liquid drainage through Plateau borders was investigated for a persistent foam (minimal loss in foam height). A macroscopic drainage study was done to identify the time frame associated with liquid drainage. Data were collected by MRI for egg white foam, a protein-stabilized foam, within this time frame and the results were compared to a basic theoretical model for liquid drainage developed for a persistent foam (2). The profiles reported here are qualitatively similar to those reported by Assink, et al (5). However, signal intensities have been corrected for relaxation effects as described by McCarthy (6). The resulting foam density data is quantitatively compared to theoretical predictions.

## THEORY

The basic model for liquid drainage used in this study was developed by Kraynik (2). This model incorporated several assumptions in order to minimize the number of parameters. The conditions are summarized as follows: 1) The transient liquid drainage through interconnecting Plateau borders is modeled for a stationary column of foam. 2) Because of crowding ($f<0.259$), gas cells in the foam are modeled as regular pentagonal dodecahedra separated by thin flat liquid films, the lamellae. 3) The initial liquid volume fraction is uniform throughout the foam. 4) The gas-liquid interfaces are considered rigid. 5) Liquid drains away from the foam at the bottom of the column.

The partial differential equation that governs drainage from a stationary foam was derived from a mass balance for the Plateau borders over a differential volume (1, 7). Flow through an individual Plateau border, which is considered to be a long channel with uniform cross section, at a given angle $\theta$ to the horizontal plane was evaluated (1, 2, 8, 9). The partial differential equation, in dimensionless form, set the sum of the rate of change of liquid volume fraction and change in liquid flux equal to zero (2).

$$\frac{\partial U}{\partial \tau} + U\frac{\partial U}{\partial \xi} = 0 \qquad (1)$$

The value of U is the volume fraction of liquid, which is the value of $f=f(z,t)$ made dimensionless by the initial volume fraction of liquid, $\langle f \rangle$. The independent variable $\xi$ is the vertical position z, normalized by the height of the foam column, H. The variable t is time made dimensionless by the characteristic time, $t_o$. The characteristic time is a function of the liquid density and viscosity, the average size of the gas cells, the initial volume fraction of liquid, the height of the foam column, and the gravitational constant.

Subject to the initial condition of uniform initial volume fraction and the boundary condition of no liquid entering the column top, the solution for the dimensionless liquid volume fraction as a function of vertical position and time is (2):

$$U = \xi / \tau \qquad \xi/\tau < 1 \qquad (2)$$

$$U = 1 \qquad \xi/\tau > 1 \qquad (3)$$

The foam density curves predicted by this model are superimposed on experimental data in the Results and Discussion section.

## MATERIALS AND METHODS

A two Tesla CSI-II multinuclear spectrometer and imaging system (General Electric) was used to acquire data. $^1H$ spectra were recorded at 85.5 Mhz using a Fourier Imaging spin-echo sequence (General Electric). One-dimensional projections were acquired by turning the phase encode gradient off. Predelay was set to five times the spin-lattice relaxation time. Spin-spin relaxation effects were corrected according to the procedure described by McCarthy (6). For one-dimensional projections 10 ms echo delay, 8 sec predelay, 10 mm slice thickness, 128 point resolution, and 2.94 ms acquisition time were used.

The foam used in this study was reconstituted egg white foam (12.5% solids). The reconstituted egg white was whipped for 2 minutes on high speed (Sunbeam mixer). A sample of egg white was placed in the cubic sample cylinder (4 cm x 4 cm x 4 cm). A plastic mesh screen (with 2 $mm^2$ openings) at the sample holder bottom contained the foam, but allowed the separated liquid to drain away.

During the 60 minute macroscopic drainage study, negligible volume loss was observed. The MRI experiment was planned for a time interval of 20 minutes to ensure evaluation of primarily liquid drainage. Magnetic resonance imaging commenced at 1 minute after the completion of the sample preparation (denoted as t = 0 min).

The following parameters for egg white foam were measured by standard methods and are used in the comparison of theory to experimental results: uniform initial liquid volume fraction, <f>=0.102; liquid density, r=1.027 $g/cm^3$; liquid viscosity, m=0.037 Poise; height of foam column, H=4.0 cm.

The final parameter, d, is an estimate of the diameter of a sphere whose volume is equal to that of the gas-filled cell. The determination of reliable air-cell size remains extremely difficult in all but the most ideal systems. In general, the best method of gas-cell size determination is by microscopic methods (10). The average bubble size for egg white has been determined by photomicrograph and reported to be 0.02 cm for a comparable whip time to this study (11). This value falls in the range reported by German et al. (12) for bubble size distribution for 5% egg white protein solution. For this bubble diameter, d is estimated to be 0.022 cm. Using the parameters above, the characteristic time, $t_o$, used in the theoretical model, was 22.1 minutes.

## RESULTS AND DISCUSSION

The Magnetic Resonance Imaging technique was performed to gain insight into the draining phenomena during the first 20 minutes of liquid drainage. Figure 1 is a one-dimensional MRI projection of foam density versus position. The curves shown are representative of the data obtained from this procedure and are similar to those given by McCarthy, et al. (13). The curves in Figure 1 are projections taken at 0 minute, 10 minutes and 20 minutes. The left side of the diagram corresponds to the bottom of the container; the right side of the diagram corresponds to the top of the container. The foam density is relatively uniform at vertical positions throughout the container at t=0 minute. As illustrated in the profiles at t=10 and 20 minutes, the liquid in the top portion of the foam has drained and the foam density in this region decreases. As the liquid drains by gravity through the Plateau borders to the bottom of the container, the foam becomes more dense. During the experimental time interval of 20 minutes, approximately 6% of the liquid drained free of the foam. This degree of liquid separation was verified by the macroscopic drainage out of the foam column which began 12 minutes after completion of whipping. However, the height of the foam remained constant, as observed at the container top (Figure 1).

Based on the MRI results, the model does not adequately predict the foam density as a function of position or time (Figure 2). The model predicts a much more rapid and dramatic decrease in foam density in the upper region of the column. In addition, the model does not predict any increase in foam density in the lower portion of the container. In other words, the physical interpretation of the model is that liquid is draining into the bottom region of the container without any increase on foam density. This would require liquid to separate from the foam immediately. However, the experimental data indicate that the dimensionless foam density approaches 1.5 prior to liquid separation.
In a recent study by Assink, et al. (5), MRI was also used to evaluate foam density as a function of time and position. The purpose of their work

was to demonstrate the feasibility of profiling foam density by MRI. They also compared their experimental results for a water/olefin surfactant system to the basic model developed by Kraynik (2). However, these researchers fit the value of the characteristic time, $t_o$, to their data. Because of this fitting process, the experimental results and model predictions appear to be in agreement. In contrast, this study evaluated the characteristic time based on the physical properties of the foam system.

The theoretical model developed by Kraynik (2) is an important contribution to the understanding of foam drainage; the model identifies several important parameters. However, experimental data obtained noninvasively by MRI qualitatively show deviation from this basic model. Future work will include modifying the model theoretically to account for surface tension and liquid holdup between the vapor cells.

ACKNOWLEDGEMENT

This project was funded in part by a grant from the University of California, Davis NMR Facility. We would like to express our appreciation to Al Wilson and Gerald Anderson for designing the experimental sample cell. In addition, we would like to acknowledge Henningsen Foods, Inc. (Omaha, NE) for their donation of dehydrated egg white.

## NOTATION

### Roman

d diameter of sphere whose volume is equal to that of the gas- filled cell
f liquid volume fraction; <f> initial
g gravitational constant
H height of foam column
$t_o$ characteristic time, $t_o = 580\ \mu H/(\rho\ g\ d^2 \langle f \rangle)$
U ratio of liquid volume fractions, f/<f>
z position from the container top

### Greek

$\mu$ liquid viscosity
$\rho$ liquid density
$\tau$ dimensionless time, $t/t_o$
$\xi$ dimensionless position, z/H

## LITERATURE CITED

1. Barber, A.D. and S. Hartland, Trans. Instn Chem. Engrs., 53, 106 (1975).

2. Kraynik, A.M., "Foam drainage," Sandia Report SAND83-0844, Sandia National Laboratories, Albuquerque, New Mexico (1983).

3. German, J.B. and M.J. McCarthy, J. Ag. and Food Chem., 37, 1321 (1989).

4. Bikerman, J.J., Foams, Springer-Verlag, New York (1973).

5. Assink, R.A., A. Caprihan, and E. Fukushima, AIChE J., 34 (12), 2077 (1988).

6. McCarthy, M.J., AIChEJ., 36 (2), 287 (1990).

7. Arpaci, V.S. and P.S. Larsen, Convection Heat Transfer, Prentice-Hall, Inc. Englewood Cliffs, NJ (1984).

8. Leonard, R.A. and R. Lemlich, AIChE J., 11 (1), 18 (1965).

9. Leonard, R.A. and R. Lemlich, AIChE J., 11 (1), 25 (1965).

10. Bee, R.D., A. Clement, and A. Prins, "Behaviour of an aerated food model," in Food Emulsions and Foams, Dickinson, E. (Ed.), The Royal Society of Chemistry, London (1987).

11. Barmore, M.A., Colorado Experiment Station, 9, 3 (1934).

12. German, J.B., T.E. O'Neill, J.E. Kinsella, JAOCS., 62 (9), 1358 (1985).

13. McCarthy, K.L., M.J. McCarthy, and J.B. German, "Modeling of foam drainage and collapse: incorporation of vertical mass distribution," presented at 5th International Congress on Engineering and Food, Cologne, Federal Republic of Germany (1989).

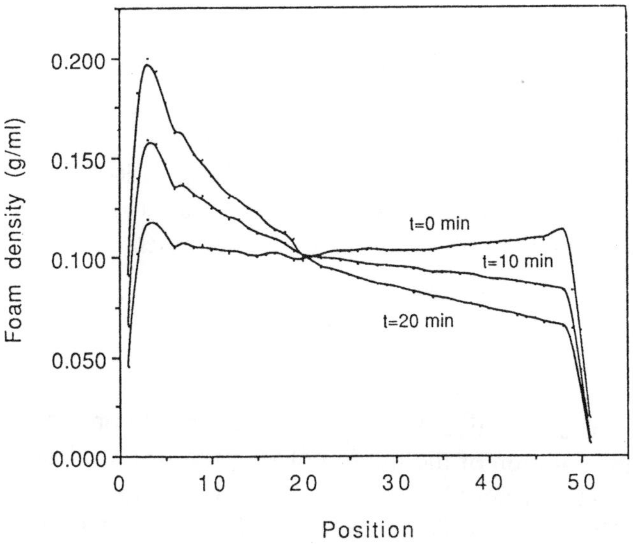


Figure 1. One-dimensional MRI projection of egg white foam density versus position. The container bottom is at 0; the container top is at 50. Projections are taken at t = 0, 10, and 20 minutes.

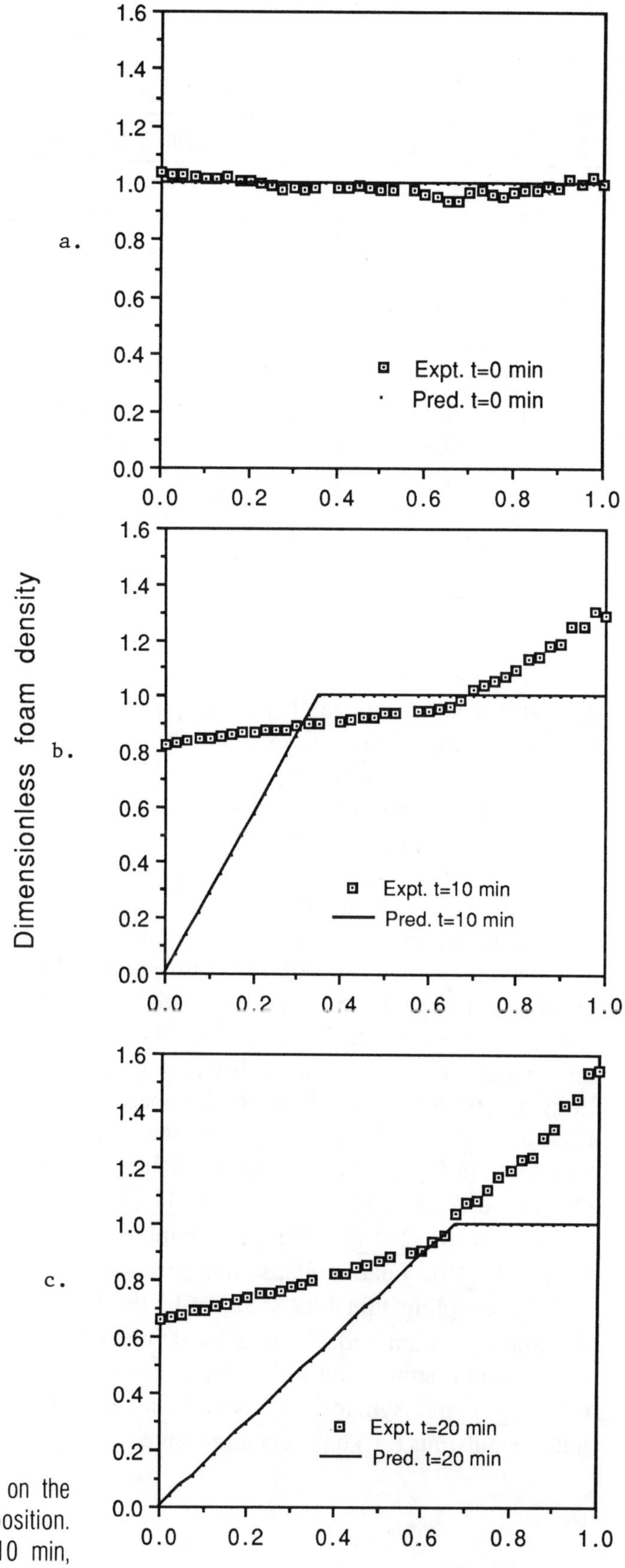


Figure 2. Model-predicted curves are superimposed on the MRI data for egg white foam density versus position. a) comparison at t = 0 min, b) comparison at t = 10 min, and c) comparison at t = 20 min.

# UNSTEADY STATE DRAINAGE OF A STANDING FOAM

Ganesan Narsimhan ■ Department of Agricultural Engineering, Purdue University, West Lafayette, IN 47907

General equations for unsteady state drainage of a static foam have been developed based on a model for the hydrodynamics of a foam bed which accounted for liquid drainage from thin films and plateau borders as well as the movement of foam-liquid interface due to the accumulation of drained liquid and assumed that the foam bed consisted of dodecahedral bubbles of the same size. Based on the timescales of film and plateau border drainage, the simplifying assumption of negligible liquid in thin films was justified. Initial liquid holdup profiles were evaluated for foam bed generated by bubbling inert gas in the form of equal sized bubbles through a liquid pool at a constant flow rate by employing the hydrodynamics of moving foam. Foam drainage was found to be sensitive to the initial liquid holdup profile and therefore to the manner in which foam is generated. Foam drainage was found to be faster and the extent of drainage larger for larger initial foam heights, larger superficial gas velocities and smaller surface tensions. The rate and the extent of drainage exhibited a maximum at an optimum bubble size when the same superficial gas velocity was employed for foam generation. For the same initial liquid holdup profiles, the rate of foam drainage was found to vary inversely with viscosity whereas the extent of drainage remained the same. When the same superficial gas velocity was employed, however, more drainage occurred for larger viscosities with the rate of drainage being insensitive to the variations in viscosity.

Foamability and foam stability are important functional properties of proteins essential in many food formulations. It is customary to infer the stability of foam from the transients of drainage and collapse by employing experimental techniques such as static drainage (1,2), surface decay (3–5), half-life of foam (6) and half-life of drainage (7). Even though, the qualitative effects of different parameters on the formation and stability of protein-stabilized foams have been elucidated through extensive experimental investigations (8), studies on quantitative prediction of foam drainage are limited. Liquid holdup in foam has been calculated accounting for (i) gravity drainage from plateau borders (1,9,10), (ii) drainage of liquid from plateau border as well as thin films (11), (iii) surface viscosity (12–14) and (iv) interbubble gas diffusion (15,16). Foam collapse has been predicted (17) by coupling thin film stability to the hydrodynamics of foam bed. General equations for the prediction of aeration ratio (18), hydraulic conductivity (19) and syneresis of foams and concentrated emulsions (20) have been presented.

The objective of the present paper is to predict the evolution of the liquid holdup profile as well as the foam-liquid interface for unsteady state drainage in a standing foam (static drainage). General equations for unsteady state drainage of liquid from thin films as well as from plateau borders are presented in the next section. Simplified equation for unsteady state drainage based only upon drainage of liquid from plateau border is then presented along with appropriate boundary conditions as well as the equation describing the rate of movement of foam-liquid interface as a result of accumulation of liquid from a draining foam. Equation describing the variation of liquid holdup with position in a rising foam is then presented along with appropriate boundary conditions in order to obtain the initial liquid holdup profile for the drainage of a standing foam. Model predictions for the evolution of the liquid holdup profile as well as the foam-liquid interface for different parameters such as surface tension, bubble size, superficial gas velocity, foam height, and viscosity are presented in the subsequent section.

## MODEL FOR DRAINAGE OF A STANDING FOAM

The dispersed bubbles of a foam, on the average, can be considered to be regular pentagonal dodecahedrons separated by thin films. Three adjacent thin films intersect in a plateau border and the continuous phase is interconnected through a network of plateau borders (21). In a static foam, the liquid in thin films drains into the neighboring plateau borders under the action of plateau border suction, followed by drainage through the network of plateau borders due to gravity. Eventually, the drained liquid accumulates at the bottom of the standing foam, thus resulting in a profile of liquid holdup and film thickness, both increasing from the top to the bottom of the foam bed.

The evolution of liquid holdup profile in a static foam is shown schematically in Figure 1. The foam bed is assumed to consist of dodecahedral bubbles of the same size. Since liquid drainage occurs along the vertical direction (direction of gravity), the foam bed can be assumed to be uniform across the bed cross section. The z coordinate is taken to be along the direction of gravity with z = 0 and z = L referring to the top of the foam and the foam-liquid interface respectively. An unsteady state material balance for the liquid in a film yields,

$$\frac{\partial}{\partial t} x_f = - V_f \tag{1}$$

where $V_f$ is the velocity of film drainage. As will be pointed out later, the velocity of film drainage would depend on various parameters such as the film thickness, surface tension, radius of curvature of plateau border, etc. The number of bubbles per unit volume N is given by,

$$N = \frac{(1 - \varepsilon)}{v} \tag{2}$$

where ε is the liquid holdup and v (= $4 \pi R^3/3$) is the volume of bubbles of radius R. The total number of plateau borders intercepted by a unit cross sectional area of the foam bed is $2 N R n_p/5$ (17), where $n_p$, the number of plateau borders per bubble, is equal to ten for dodecahedral arrangement (14). For randomly oriented plateau borders, the average volumetric flow rate per plateau border is $2 a_p u/3$ (17), where $a_p$ is the area of cross section of a plateau border and u is the velocity of gravity drainage. An unsteady state material balance for the liquid in the plateau borders over a volume element between z and z + dz yields,

$$\frac{\partial}{\partial t} (N n_p a_p \ell) = - \frac{4}{15} \frac{\partial}{\partial z} (N n_p a_p u R) + N n_f A_f V_f \tag{3}$$

where $A_f$ is the area of a film, $n_f$ (= 12) is the number of films per bubble and $\ell$ is the length of a plateau border. In the above equation, the first term refers to the accumulation whereas the second and third terms refer to the change in the volume of liquid in plateau borders due to gravity drainage and drainage from films respectively. In order to obtain the evolution of the liquid holdup profile, the two coupled partial differential equations (Equations (1) and (3)) have to be solved with the appropriate initial and boundary conditions and the liquid holdup ε is related to the film thickness $x_f$ and the plateau border area $a_p$ by

$$\varepsilon = N n_f A_f x_f + N n_p a_p \ell \tag{4}$$

In order to solve Equations (1) and (3), however, one should have knowledge of u and $V_f$. The evaluation of these quantities are discussed in the subsequent sections.

### Velocity of Gravity Drainage Through Plateau Border

The velocity of drainage u through a plateau border can be obtained by solving the Navier-Stokes equations for the plateau border geometry with the appropriate boundary conditions which account for the surface mobility. In order to simplify the analysis, the actual shape of plateau border cross section has been idealized by various geometries (1,12,13,22). In what follows, the solution given by Desai and Kumar (22), who have approximated the cross section by an equilateral triangle, is modified. Their expression for the velocity of drainage u is given by:

$$u = \frac{c_v\, a_p}{20\sqrt{3}\,\mu}\,\rho\, g \tag{5}$$

where $\rho$ is the density of the liquid, g is the acceleration due to gravity, $\mu$ is the viscosity of the liquid and $c_v$, the velocity coefficient, is defined as the ratio of the average velocity through the plateau border to that for infinite surface viscosity. In their analysis, Desai and Kumar (22) have expressed the general solution in terms of pressure gradient which, in their subsequent calculation for the liquid holdup profile (14), is assumed to be that due to gravity. In other words, they have assumed that the force per unit volume responsible for flow of liquid in the plateau border is $\rho$ g. However, there will, in general, be an additional force other than gravity because of the gradient of pressure within the liquid in the plateau border along the z direction (vertical direction) resulting from the variation of plateau border suction. The force per unit volume responsible for the flow of liquid in the plateau border should therefore be ($\rho$ g - dp/dz) (20). The pressure within the plateau border p can be related to the pressure within the gas bubbles $p_d$ through

$$p = p_d + p_c\ , \tag{6}$$

where the plateau border suction $p_c$ is given by

$$p_c = \frac{-\sigma}{r}\ , \tag{7}$$

$\sigma$ and r being the surface tension and the radius of curvature of the plateau border respectively. Since the dispersed phase (the gas bubbles) can be assumed to be in hydrostatic equilibrium, the variation of the dispersed phase pressure would be given by,

$$\frac{dp_d}{dz} = \rho_d\, g \tag{8}$$

$\rho_d$ being the density of the dispersed phase. Combining Equations (6), (7) and (8), one obtains,

$$\frac{dp}{dz} = \rho_d\, g - \sigma \frac{d}{dz}\left(\frac{1}{r}\right) \tag{9}$$

The expression for the velocity of drainage u through the plateau border in terms of the modified driving force is therefore given by:

$$u = \frac{c_v\, a_p}{20\sqrt{3}\,\mu}\left[(\rho - \rho_d)\, g + \sigma \frac{d}{dz}\left(\frac{1}{r}\right)\right] \tag{10}$$

Since $\rho_d << \rho$, the above equation can be approximated by

$$u = \frac{c_v\, a_p}{20\sqrt{3}\,\mu}\left[\rho\, g + \sigma \frac{d}{dz}\left(\frac{1}{r}\right)\right] \tag{11}$$

From geometric considerations (12), the radius of curvature of the plateau border r can be related to the area of the plateau border $a_p$ and the film thickness $x_f$ by

$$r = \frac{-1.732x_f + \left[(1.732x_f)^2 - 0.644(0.433\, x_f^2 - a_p)\right]^{1/2}}{0.322} \tag{12}$$

The velocity coefficient $c_v$ is a function of the inverse of the dimensionless surface viscosity $\gamma$ defined as (22)

$$\gamma = \frac{0.4387\mu a_p^{1/2}}{\mu_s} \tag{13}$$

where $\mu_s$ is the surface viscosity. The functional dependence of $c_v$ on $\gamma$ is expressed through a spline fit as (14)

$$c_v = b_{i0} + b_{i1}(\gamma - \gamma_i) + b_{i2}(\gamma - \gamma_i)^2$$

$$+ b_{i3}(\gamma - \gamma_i)^3 \tag{14}$$

where $\gamma_i \leq \gamma \leq \gamma_{i+1}$, i=1, 2, ...5. The values of the coefficients $\{b_{ij}, j=0\text{–}3\}$ and the constants $\gamma_i$ can be found elsewhere (14).

## Velocity of Drainage of a Foam Film

The velocity of drainage of a circular plane parallel film of radius $R_f$ and thickness $x_f$ is given by the Reynolds equation,

$$V_f = \frac{2\Delta p x_f^3}{3\mu R_f^2} \tag{15}$$

if the film surface is immobile, i.e. the surface viscosity is very large. For a dodecahedral arrangement, the radius of the film $R_f$ can be related to the radius of the bubble R through geometric considerations by (14)

$$R_f = 0.808\ R \tag{16}$$

The pressure drop $\Delta p$ responsible for the thinning of the film is the sum of the contributions from plateau border suction and the disjoining pressure $\Pi$ and is given by

$$\Delta p = \sigma/r - \Pi \tag{17}$$

It is worth noting that the disjoining pressure $\Pi$ due to van der Waals, double layer and steric interactions would become important only when the film thickness is of the order of a thousand angstroms or smaller. When the film surface is mobile, i.e. the surface viscosity is not very large, the rate of film thinning would be larger than that predicted by the Reynolds equation and can be obtained by solving the Navier-Stokes equation with appropriate boundary conditions accounting for the mobility of the film surface as well as the effects of surface and bulk diffusion of the surfactant (23).

## Simplified Equations for Drainage

In order to simplify the analysis of foam drainage, the timescales of film and plateau border drainage are compared. A typical comparison is shown in Figure 2 as a plot of the timescales versus the fraction of liquid in films $\phi_f$ for different liquid holdups. The timescale of film drainage $\tau_f$ is found to be a strong function of $\phi_f$ (and therefore $x_f$) increasing by several orders of magnitude as the film thickness decreases (refer to Figure 2). The timescale of plateau border drainage $\tau_p$, however, is fairly insensitive to the variations in $\phi_f$. As a result, $\tau_f$ is much smaller than $\tau_p$ for thick films and much larger for thin films the crossover film thickness being dependent on the liquid holdup. Therefore, thick films would continue to drain extremely rapidly into the neighboring plateau borders until the film thickness decreases below the crossover thickness beyond which the rate of film drainage as well as the fraction of liquid in films become negligible. The relaxation time for this initial rapid drainage from thick films is expected to be extremely small so that the velocity of film drainage can be neglected in the unsteady state equations for foam drainage. Consequently, the unsteady state drainage of a standing foam can be described by a single equation corresponding to the gravity drainage of liquid in the plateau border given by:

$$\frac{\partial}{\partial t}(N n_p a_p \ell) = -\frac{4}{15}\frac{\partial}{\partial z}(N n_p a_p u R) \tag{18}$$

where the velocity of gravity drainage u is given by Equation (11). The simplified expressions for the liquid holdup $\varepsilon$ and the radius of curvature of the plateau border r are given by:

$$\varepsilon = N n_p a_p \ell \tag{19}$$

and

$$r = \alpha a_p^{1/2} \tag{20}$$

where $\alpha$ is a constant ($= \sqrt{0.644}/0.322$). From Equations (1), (19) and (20) the area of the plateau border $a_p$ and the radius of curvature r can be related to the liquid holdup $\varepsilon$ as:

$$a_p = \left(\frac{v}{n_p \ell}\right)\left(\frac{\varepsilon}{1-\varepsilon}\right) \tag{21}$$

and

$$r = \alpha \left(\frac{v}{n_p \ell}\right)^{1/2} \left(\frac{\varepsilon}{1-\varepsilon}\right)^{1/2} \qquad (22)$$

Using Equations (11) and (19), the unsteady state drainage Equation (18) can be recast in terms of the liquid holdup ε to give,

$$\frac{\partial \varepsilon}{\partial t} = \frac{-\rho g}{75\sqrt{3}\mu\delta} \left(\frac{v}{n_p \ell}\right) \frac{\partial}{\partial z}$$

$$\left[\frac{\varepsilon^2}{(1-\varepsilon)} \left\{1 - \frac{\sigma}{2\alpha\rho g} \left(\frac{n_p \ell}{v}\right)^{\frac{3}{2}} \left(\frac{1-\varepsilon}{\varepsilon}\right)^{\frac{3}{2}}\right\}\right] \qquad (23)$$

Boundary Conditions. If the dispersed phase bubbles are sufficiently small, then the capillary pressure inside a (spherical) bubble would far exceed the variation of hydrostatic pressure over a bubble diameter. This would be valid if the bubble radius R satisfies the following inequality,

$$R << \left[\frac{\sigma}{(\rho - \rho_d) g}\right]^{1/2} \qquad (24)$$

In such a case, the bubbles at the bottom of the foam bed, i.e. at the foam-liquid interface, would essentially be spherical (24) so that the continuous phase liquid holdup at the interface $\varepsilon_0$ can be taken to be that corresponding to closed pack spheres. Therefore,

$$\varepsilon = 0.26 \qquad \text{at} \quad z = L(t) \qquad (25)$$

where L(t) is the length of the foam bed at time t. It is to be noted that the height of the foam bed would decrease with time because of the continuous drainage and accumulation of the liquid at the bottom. In other words, the foam-liquid interface boundary would be moving with respect to time. The actual location of the boundary at different times is to be determined through the material balance of liquid. Since there is no input of liquid at the top of the foam bed (z=0), the velocity of gravity drainage of the plateau border should be zero at that location. This condition would be satisfied if the net driving force for plateau border drainage at the top of the foam bed is zero. The second boundary condition is, therefore, given by:

$$u = 0 \qquad \text{at} \quad z = 0 \qquad (26)$$

which implies that

$$\sigma \frac{\partial}{\partial z}\left[\frac{1}{r}\right] = -\rho g \qquad \text{at} \quad z = 0 \qquad (27)$$

A material balance for the liquid in the foam yields,

$$\int_0^{L(t)} \varepsilon(z,t)\, dz + L_o - L(t) = \int_0^{L_o} \varepsilon_0(z)\, dz \qquad (28)$$

where $\varepsilon_0(z)$ is the initial liquid holdup profile, $L_o$ and L refer to the foam heights at times zero and t respectively. The first term in the above equation refers to the total amount of liquid contained in the draining foam at time t whereas the right hand side of the equation refers to the initial amount of liquid contained in the foam. Differentiating the above equation and combining with the unsteady state drainage Equation (18) and the boundary conditions (24) and (25) one obtains the following expression for the rate of movement of the foam-liquid interface,

$$\frac{dL}{dt} = -\frac{1}{0.76}\frac{4}{15} N n_p a_p u R \Big|_{z=L(t)} \qquad (29)$$

In order to obtain the evolution of the liquid holdup profile and the foam-liquid interface for a standing foam, the unsteady state balance Equation (18) has to be solved along with the boundary conditions (25) and (26) as well as the Equation (29) describing the movement of the interface. Of course, the initial distribution of the liquid holdup profile should also be known. This initial condition would depend on the manner in which the foam is generated. The evaluation of the initial liquid holdup profile when the foam is generated by bubbling inert gas through a pool of liquid is discussed in the next section.

Initial Condition. In the present analysis, it is assumed that the foam is generated by bubbling an inert gas through a liquid pool. As the inert gas is sparged into the liquid pool, the inert gas bubbles rise up to form a foam at the top of the liquid pool. The foam moves up the column entraining some liquid from the liquid pool. The entrained liquid in the plateau borders drains under the action of gravity through the moving foam thus resulting in a profile of the liquid holdup (which increases from the top to the bottom of the moving foam). For foam bed consisting of bubbles of the same volume v, the number of bubbles that flow per unit area of the foam bed per unit time $\eta$ is given by,

$$\eta = \frac{G}{v} , \tag{30}$$

where G is the superficial gas velocity. For sufficiently small superficial gas velocities, the rates of entrainment and drainage of liquid can be assumed to be in quasi-steady state. A material balance for the liquid in the plateau border over a volume element between z and z + dz would, therefore, yield:

$$\frac{d}{dz} ( \eta a_p n_p \ell ) - \frac{4}{15} \frac{d}{dz} ( N n_p a_p u R ) = 0 \tag{31}$$

Using Equations (19) and (30), the above equation can be recast in terms of liquid holdup $\varepsilon$ as

$$G \frac{d}{dz} \left[ \frac{\varepsilon}{1-\varepsilon} \right] - \frac{4}{15} \frac{d}{dz} \left[ \frac{\varepsilon u}{\delta} \right] = 0 \tag{32}$$

As the foam moves up, the height of the foam bed would increase. At the foam-liquid pool interface, the liquid holdup would be that corresponding to closed packed spheres. If z = 0 refers to the top of the moving foam,

$$\varepsilon = 0.26 \qquad \text{at} \quad z = L \tag{33}$$

where L is the instantaneous height of the moving foam. The velocity of the top of the moving foam would be $G/(1-\varepsilon)$. A material balance for the liquid with respect to an observer moving with the top of the foam would yield,

$$u = 0 \qquad \text{at} \quad z = 0 \tag{34}$$

In order to generate a standing foam of initial height $L_o$, the inert gas is to be shut off at the instant the height of the moving foam reaches $L_o$ so that the initial liquid holdup profile can be obtained by solving equation (32) with the boundary conditions (33) and (34).

## EVOLUTION OF FOAM–LIQUID INTERFACE AND LIQUID HOLDUP PROFILE FOR A STATIC FOAM

The evolution of liquid holdup profile was calculated by solving Equation (18) with the boundary conditions (25) and (27). The location of the foam-liquid interface was updated using Equation (29). The initial liquid holdup profile for a static foam was obtained by solving Equation (32) with the boundary conditions (33) and (34). Transients of liquid holdup profile were calculated to investigate the effects of bubble size, superficial gas velocity employed for foam generation, viscosity, surface viscosity and the initial foam height. The complete results of the calculations can be found elsewhere (25).

Typical evolution of liquid holdup is shown in Figure 3. The liquid holdup at the foam-liquid interface remains constant at 0.26 whereas the liquid holdup decreases with time in the rest of the foam bed thus making the profile more steeper at larger times. Moreover, the height of the foam bed decreases with time because of the accumulation of drained liquid at the bottom. Even though the initial change in the liquid holdup profile is rapid, the subsequent rates of drainage become smaller because of smaller driving force for drainage as a result of larger gradients of plateau border suction (which opposes gravity). Eventually, gravity is exactly counterbalanced by the gradient of plateau border suction after which no further liquid drainage occurs and the foam bed reaches equilibrium.

Figure 4 shows the effect of bubble size on the transients of foam drainage as a plot of dimensionless height of drained liquid (defined as $(L_0 - L)/L_0$) versus time. The rate and the extent of drainage are found to increase with bubble size for smaller bubbles, attain a maximum and decrease with bubble size for larger bubbles. For the same liquid holdup, the rate of plateau border drainage tends to increase with bubble size because of the increase in the area of cross section of the plateau border (due to smaller surface area per unit volume of the foam). On the other hand, the initial liquid holdup profile tends to be steeper for larger bubbles (see Figure 5) because of greater relative rates of plateau border drainage compared to entrainment during foam formation. As a result, the gradient of plateau border (which opposes gravity), being larger for larger bubbles, tends to slow down the rate of plateau border drainage for larger bubbles. The existence of an optimum bubble size (Figure 4) can be attributed to the above two opposing effects.

The rate as well as the extent of foam drainage are larger when larger superficial gas velocities are employed for the formation of the standing foam (see Figure 6). The initial liquid holdup profile is steeper at smaller superficial gas velocities because of smaller relative rates of entrainment compared to drainage. Consequently, the gradient of plateau border suction is steeper for smaller values of G thus resulting in smaller rates of drainage.

Larger rate and the extent of drainage are found for larger initial foam heights as can be seen from Figure 7. More liquid is entrained with smaller gradient of initial liquid holdup for taller foam beds. Consequently, taller foam beds exhibit faster rates of drainage because of smaller gradient of plateau border suction opposing gravity. Also, it takes longer for the liquid holdup profile to propagate through a taller foam bed and therefore longer for the foam bed to attain equilibrium.

In order to investigate the effect of viscosity on foam drainage, two sets of calculations were performed. In the first, the superficial gas velocity was varied during foam formation for solutions of different viscosity in order to produce foams of the same initial liquid holdup profile. Since the rate of plateau border drainage is inversely proportional to viscosity, the rate of entrainment (and therefore the superficial gas velocity) has to be decreased in proportion to the increase in the viscosity in order to maintain the same ratio of the rates of entrainment and drainage so as to produce foams of the same initial liquid holdup profiles. As expected, an increase in the viscosity decreases the rate of drainage as can be seen from Figure 8. However, the extent of drainage is found to be independent of viscosity because of the same initial liquid holdup profile. In the second set of calculations, the effect of viscosity on the rate and the extent of drainage were compared when the same superficial gas velocity was maintained during foam formation. The rate of entrainment of liquid compared to the rate of drainage during foam formation was higher for higher viscosity solutions thus resulting in a more uniform initial liquid holdup profile. As can be seen from Figure 9, the effect of viscosity on drainage is confounded by the effect of initial liquid holdup profile. The rate of drainage is not influenced strongly by viscosity. However, the extent of drainage is found to be more for more viscous solutions. Such a behavior can be explained in terms of the difference in the initial liquid holdup profiles. For more viscous solution, the contribution to the driving force from the gradient of the plateau border suction (which opposes gravity) is smaller because of more uniform initial liquid holdup profile. As a result, the net driving force for plateau border drainage (gravity minus the gradient of plateau border suction) is larger. However, the rate of drainage for a specified driving force would be smaller because of higher viscosity of the solution. Because of these opposing effects, the rate of drainage is found to be fairly insensitive to the viscosity as can be seen from Figure 9. For more viscous solutions, the initial liquid holdup profile, being less steep than that for less viscous solutions, would be farther from the equilibrium liquid holdup profile since the equilibrium liquid holdup profile is independent of viscosity. Consequently, (i) more drainage has to occur before the foam bed reaches equilibrium and (ii) drainage would take longer (because of larger time scale of foam drainage) for liquid of higher viscosity.

Drainage is found to be much faster and the extent of drainage larger for smaller values of surface tension (see Figure 10). At smaller surface tensions, the driving force due to the gradient of plateau border suction (which opposes gravity) is smaller and consequently, the rate of foam drainage is larger. The effect is found to be quite pronounced for a small bubble size of 0.02 cm (Figure 10). It is to be noted, however, that the effect of surface tension $\sigma$ on foam drainage would be less pronounced for larger bubble sizes because of smaller contribution of the driving force due to the gradient of plateau border suction to plateau border drainage compared to gravity.

## CONCLUSIONS

General equations for unsteady state drainage of a static foam have been developed based on a model for the hydrodynamics of the foam bed which accounted for liquid drainage from thin films and plateau borders and assumed dodecahedral bubbles of the same size. Simplified equations for the evolution of liquid holdup profile were presented based on the timescales of film and plateau border drainage. Foam bed is assumed to be generated by bubbling an inert gas through a liquid pool in the form of bubbles of the same size at constant superficial gas velocity. The initial liquid holdup profiles were evaluated from the hydrodynamics of moving foam. Drainage was found to be faster and the extent of drainage larger for smaller surface tensions, larger superficial gas velocities and for taller foam beds. When the foam bed of different bubble sizes was generated using the same superficial gas velocity, an optimum bubble size was found at which the rate as well as the extent of drainage exhibited a maximum. For the same initial liquid holdup profile, the extent of drainage was found to be independent of viscosity whereas the rate of drainage varied inversely with viscosity. When the same superficial gas velocity was maintained for foam generation, however, the rates of foam drainage were not influenced very much by viscosity with the extent of foam drainage being larger for higher viscosity. Foam drainage was found to be very sensitive to the initial liquid holdup profile and therefore to the manner in which foam is generated. It is, therefore, important to standardize the method of foam generation if foam drainage is to be employed as a diagnostic test to infer its stability.

## LITERATURE CITED


1. Miles, G.D., L. Sheklovsky, and T. Ross, J. Phys. Chem., 49, 93 (1945).
2. Brady, A.P., and S. Ross, J. Phys. Chem., 66, 1348 (1944).
3. Ross, S., and M.J. Cutillas, J. Phys. Chem., 59, 863 (1955).
4. Nishioka, G., and S. Ross, J. Colloid Int. Sci., 81, 1 (1981).
5. Monsalve, A., and R.S. Schechter, J. Colloid Int. Sci., 97, 327 (1984).
6. Sharma, M.K., D.O. Shah, and W.E. Brigham, A.I.Ch.E. J., 321, 222 (1985).
7. Rand, P.B., and A.M. Kraynik, Soc. Pet. Eng. J., 23, 152 (1983).
8. Halling, P.C., Crit. Rev. Food Sci. and Nutrition, 15, 155 (1981).
9. Jacobi, W.H., K.E. Woodcock, and C.S. Grove, Ind. Eng. Chem., 48, 9046 (1956).
10. Steiner, L., R. Hunkeler, and S. Hartland, Trans. Inst. Chem. Eng., 55, 153 (1977).
11. Hartland, S., and A.D. Barbaer, Trans. Inst. Chem. Eng., 52, 43 (1974).
12. Leonard, R.A., and R. Lemlich, A.I.Ch.E. J., 11, 18 (1965).
13. Haas, P.A., and H.F. Johnson, Ind. Eng. Chem. Fundl., 6, 225 (1967).
14. Desai, D., and R. Kumar, Chem. Eng. Sci., 38, 1525 (1983).
15. Narsimhan, G., and E. Ruckenstein, Langmuir, 2, 494 (1986).

16. Lemlich, R., Ind. Eng. Chem. Fundl., 17, 89 (1978).

17. Narsimhan, G., and E. Ruckenstein, Langmuir, 2, 230 (1986).

18. Krotov, V.V., Colloid J. USSR (Engl. Transl.), 42, 903 (1980).

19. Krotov, V.V., ibid, 42, 912 (1980).

20. Krotov, V.V., ibid, 43, 33 (1981).

21. Bikerman, J.J., Foams; Springer-Verlag; New York, 1973.

22. Desai, D., and R. Kumar, Chem. Eng. Sci., 37, 1361 (1982).

23. Ivanov, I.B., and D.S. Dimitrov, Colloid Polym. Sci., 252, 982 (1974).

24. Princen, H.M., Langmuir, 2, 519 (1986).

25. Narsimhan, G., J. Food Eng., (submitted).


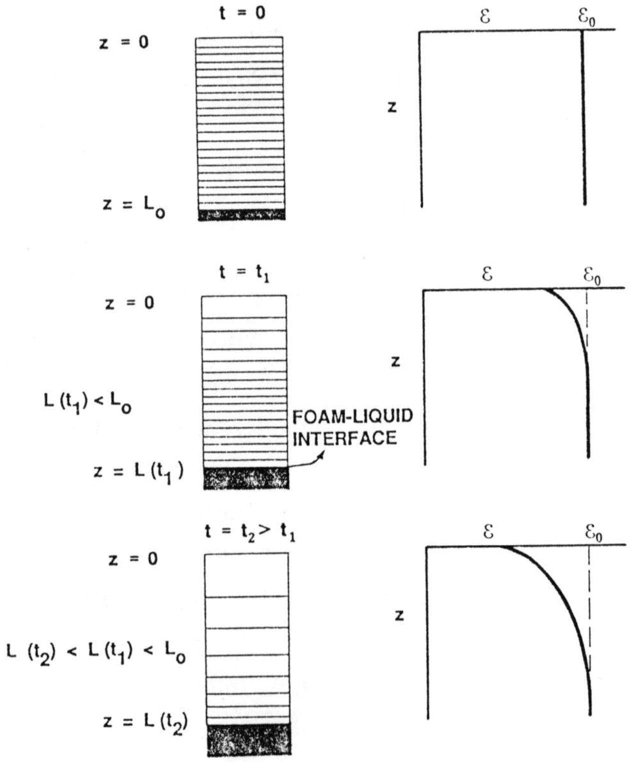


Figure 1. Schematic diagram describing the evolution of liquid holdup profile in a static foam bed. For illustrative purposes, the initial liquid holdup profile is shown to be uniform.

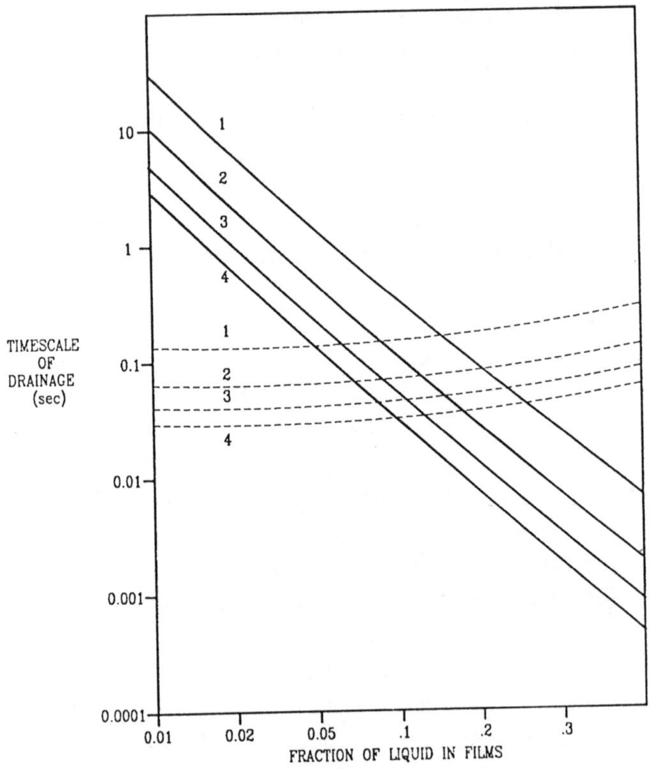


Figure 2. Plot of timescales of drainage versus fraction of liquid in films for different liquid holdups for R = 0.075 cm, $\sigma$ = 50 dynes/cm and $\mu$ = 0.01 P. Solid curves refer to the timescale of film drainage and dotted curves refer to the timescale of plateau border drainage. Liquid holdups corresponding to different curves are: curve 1 - 0.05; curve 2 - 0.10; curve 3 - 0.15 and curve 4 - 0.20.

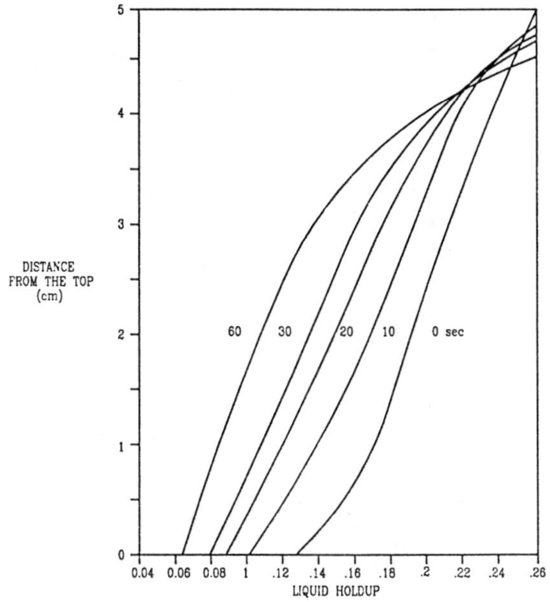


Figure 3. Evolution of liquid holdup profile in a draining foam for $L_o$ = 5 cm, R = 0.02 cm, G = 0.1 cm/sec, $\sigma$ = 50 dynes/cm, $\mu$ = 0.01 P and immobile gas-liquid interface.

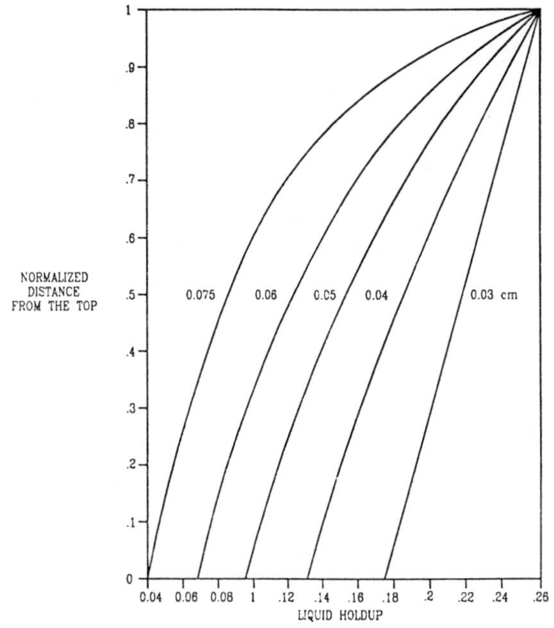


Figure 5. Initial liquid holdup profiles for different bubble sizes. Other conditions are the same as those listed in Fig. 4.

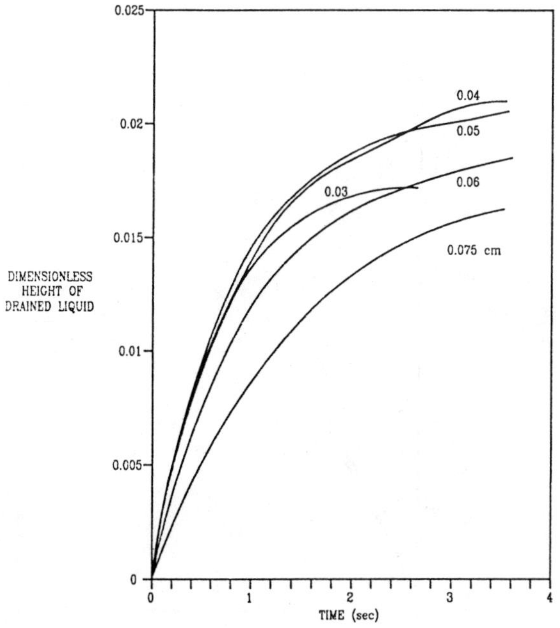


Figure 4. Plot of dimensionless height of drained liquid versus time for different bubble sizes for $L_o^* = \frac{L_o}{R} = 20$, $\sigma$ = 50 dynes/cm, $\mu$ = 0.01P, G = 0.1 cm/sec and immobile gas-liquid interface.

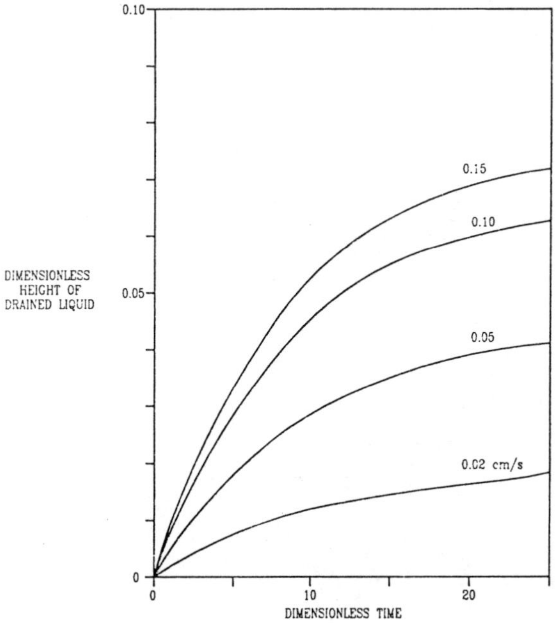


Figure 6. Plot of dimensionless height of drained liquid versus dimensionless time for different superficial gas velocities employed for foam generation for R = 0.02 cm, $L_o^* = \frac{L_o}{R} = 100$, $\sigma$ = 50 dynes/cm, $\mu$ = 0.01 P and immobile gas-liquid interface. The dimensionless time is defined as $t^* = t/\tau$, where $\tau = (n_p l/v)\,(R\mu\delta/9.253 \times 10^{-4}\,\rho g)$.

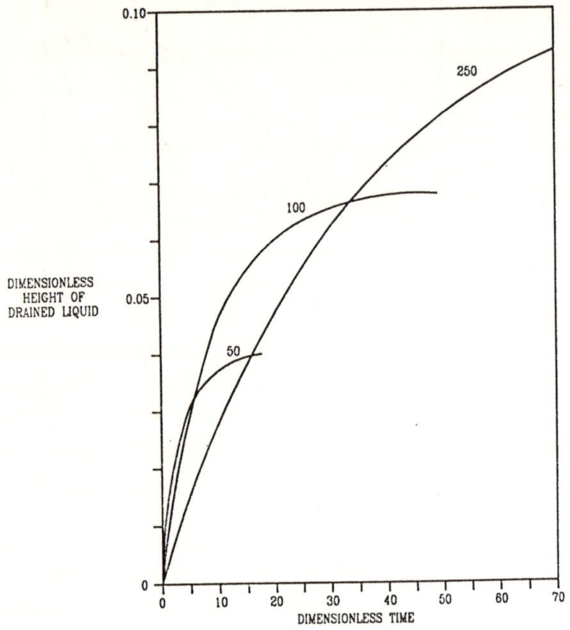


Figure 7. Plot of dimensionless height of drained liquid versus dimensionless time for different initial dimensionless foam heights for R = 0.02 cm, G = 0.1 cm/sec, $\sigma$ = 50 dynes/cm, $\mu$ = 0.01 P and immobile gas-liquid interface.

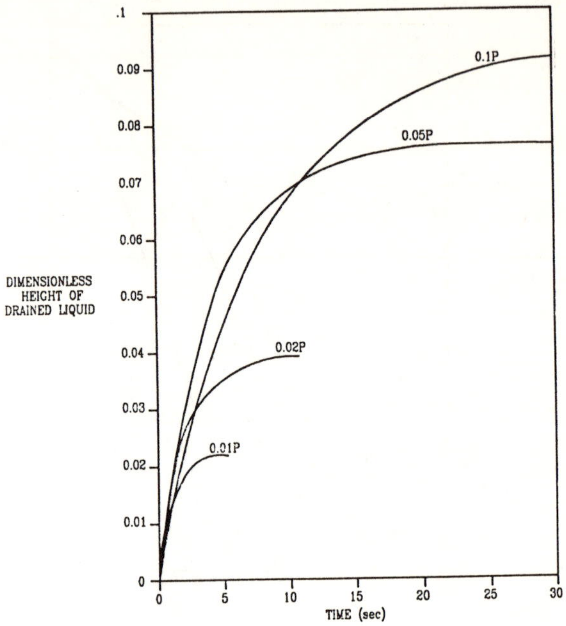


Figure 9. Plot of dimensionless height of drained liquid versus time for different viscosities for R = 0.05 cm, G = 0.1 cm/sec, $L_o^* = \frac{L_o}{R} = 20$, $\sigma$ = 50 dynes/cm and immobile gas-liquid interface.

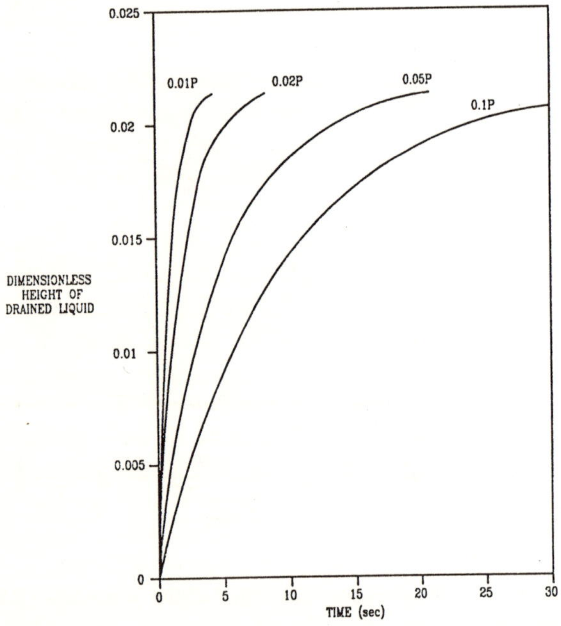


Figure 8. Plot of dimensionless height of drained liquid versus time for different viscosities for the same initial liquid holdup profile for R = 0.05 cm, $L_o^* = \frac{L_o}{R} = 20$, $\sigma$ = 50 dynes/cm and immobile gas-liquid interface.

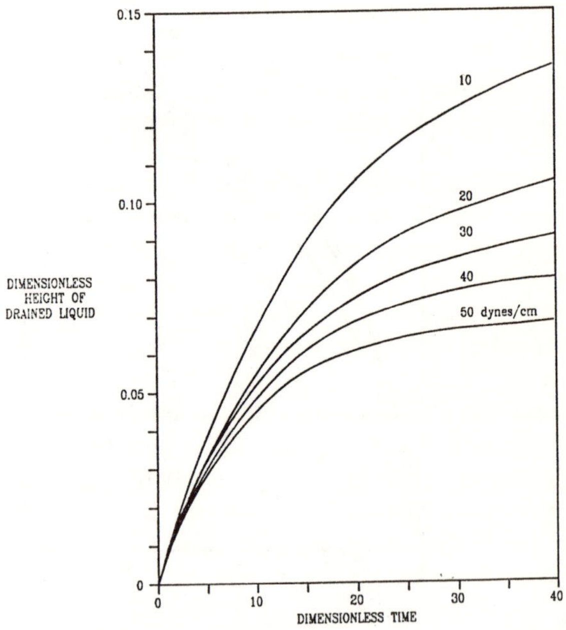


Figure 10. Plot of dimensionless height of drained liquid versus dimensionless time for different surface tensions for R = 0.02 cm, G = 0.1 cm/sec, $L_o^* = \frac{L_o}{R} = 100$, $\mu$ = 0.01 P and immobile gas-liquid interface.

# A STUDY ON THE FOAMING AND EMULSIFYING PROPERTIES OF WHEY PROTEIN HYDROLYSATES

J. Lakkis and R. Villota ■ Dept. of Food Science, University of Illinois, 382 D, A.E.S. Bldg.
1304 W. Pennsylvania Ave., Urbana, IL 61801

The foaming and emulsifying properties of three types of whey protein hydrolysates have been studied and correlated with their molecular structure. The hydrolysates studied were chymotrypsin-treated, pepsin-treated, and trypsin-treated whey protein isolates (WPI). Peptic samples showed the greatest foam expansion followed by tryptic and chymotryptic ones, while stability of foams exhibited the reverse order. The electrophoretic profile revealed that pepsin was very effective in hydrolysing WPI by means of the production of very low molecular weight peptides and by a drastic destruction of the $\beta$-lactoglobulin fraction of WPI. Fluorescence spectra showed that chymotrypsin was the most effective in changing the secondary structure of WPI by means of its greater red shift (equivalent to that of 8 M urea). Results suggested that no direct correlation could be established between the secondary structure of the proteins and their foaming capacity, though denaturation was necessary for enhancing molecular flexibility. A strong correlation was found between WPI foaming properties and its surface activity as determined from their surface tension and partition hydrophobicity. Reduction and further blocking of the intermolecular disulfide bonds was found to result in enhanced foam expansion but lower foam stability. This could be due to the possible destruction of the quaternary structure of the multi-unit protein which provides stability to the molecule. Addition of 5% native WPI to the hydrolysates had a synergistic effect on their foaming ability by means of enhancing both the ability to expand provided by the low molecular weight hydrolysates and the stability due to the undisturbed quaternary structure of the native WPI. Emulsifying properties of WPI were found to improve as shown from the smaller-sized oil droplets. Scanning electron microscopic examination of the prepared emulsions showed that chymotryptic hydrolysates were the most suitable for emulsion preparation.

Foams and emulsions are similar in theory, though in practice their behavior can be quite different. Both are dispersions of a hydrophobic fluid in an aqueous phase stabilized by suitable surface active substances (Dickinson, 1986). In foams, the disperse phase (air) is relatively soluble in the aqueous phase. This aspect is facilitated by the fact that gas molecules are small, so their diffusion rate into the water layers is rapid. In emulsions, however, the oil phase is not likely to be soluble in water, thus it forms spherical inclusions of varying sizes inside the emulsifying system.

Traditional surface-active agents include a wide array of mono- and di-glycerides, sorbitols and their esters. However, the incorporation of these ingredients into food formulations has suffered from major organoleptic and even functional drawbacks (Mansvelt, 1979). The development of waxy, bitter and painty tastes are just a few of these undesirable aspects. In addition, some of these ingredients, esters of sorbitol for example, could not find wide applicability in complex food systems due to their limited coverage up to monolayer systems. Thus, their utilization is restricted to simple systems such as fruit juices and other drinks.

Proteins, due to the absence of these problems, added to their natural prevalence in foods, are currently considered the most adequate food colloids with versatile potential. Proteins are amphiphilic polymers, which in the presence of a sufficient amount of energy (homogenization, whipping) can orient themselves at interfaces (oil/water or air/water) to stabilize foams and emulsions (Cumper, 1953; Graham and Phillips, 1976; Halling, 1981; Kinsella, 1981). During this process a new thermodynamic equilibrium is reached. Thus, the hydrophobic segments or loops spread towards the apolar phase, and charged segments tend to occupy the aqueous phase. For highly functional proteins, this process is usually perfected by means of some chain-chain interactions, hence providing adequate mechanical strength to the protein films just formed and stabilizing their structure.

The functionality of proteins, however, is known to vary extremely even with the slightest changes in their immediate environment. Food proteins, in addition, are made up of a heterogenous assembly of subunits with different chemical and/or structural nature. This makes the prediction of their potential difficult. Despite these facts, in certain instances, protein heterogeneity has been found to be an advantage. Graham and Phillips (1979) have reported that the composite-mixture of structurally different proteins in egg-albumen behaves cooperatively towards their foaming abilities by means of carrying out different functions. The globulins facilitate foam formation while the ovomucin-lysozyme complex (rigid but flexible) provides film stability. Similar predictions were offered by Johnson and Zabik (1981), although their explanation was based on other aspects of this multi-component system, i.e. electrostatic interactions. At the natural pH of fresh eggs (pH 7.0 to 8.0), lysozyme (I.Pt. 10.7) is positively charged and can interact electrostatically with the negatively charged fraction. Consistent with this argument, are the observations of De Vilbiss et al. (1974) that whey proteins, being a simple system, are less satisfactory for foaming, since a single or possibly two components must carry out most of the functions necessary for flexible and firm film formation.

Most water-soluble proteins are known to carry some inherent surface-activity, however, their usefulness in real food systems is governed by other equally essential properties. Molecular flexibility, ease of spreading at interfaces with a substantial decrease of surface tension, hydrophobicity and adequate rheological stability have been considered the most important conditions necessary for the successful incorporation of a given protein into foams and emulsions.

Hydrophobic interactions are usually considered a major aspect in stabilizing the three-dimensional structure of the protein. Birdi (1973) found a good relationship between the polar/apolar ratio of amino acids in a protein and its molecular flexibility. Proteins with ratios as low as 1 to 3 (polar to apolar) were found to unfold completely at interfaces with the more hydrocarbon-like residues oriented away from the water sites. Similar results were obtained by Kato and Nakai (1979) for a wide array of proteins including bovine serum albumin (BSA), $\beta$-lactoglobulin, k-casein and ovalbumin. Mitchell (1986) reported on the ability of hydrophobic derivatives of casein, prepared by acylation, to stabilize the lamellae in foams and in retarding film collapse. Horiuchi et al. (1978) using five different proteins, showed that the foaming power decreased in the order, gluten, egg-albumen, soybean, casein and gelatin. The same order was also found for their hydrophobicity as determined by a dye-binding technique and for surface activity as estimated from their sorption isotherms.

Surface tension and its correlation to the foaming and emulsifying properties of food proteins have been thoroughly investigated. Considerable improvement in the functional properties of vegetable proteins accompanied by a favorable depression in their surface tension was obtained upon acylation (Fox et al., 1982). Similar results were offered by Muschiolik et al. (1987) for field bean proteins which were found to possess two of the most important physico-chemical characteristics for effective emulsion formation and stabilization: low interfacial tension and high surface viscosity, both at the oil/water interface.

Rheological properties of protein films can contribute to their overall stability in various ways. Mechanisms such as reduction of the rate of drainage of water from the aqueous lamellae, enhanced cohesive forces within the two surface layers, and stabilization of air cells against disproportionation have been described as the most important.

Rheological properties of protein films are usually evaluated in terms of their preserved tertiary structure, molecular weight, surface viscosity or others. These attributes have been considered highly detrimental to the resistance of the film against shear deformation.

Saruno (1983) upon fractionating beer foams has shown that a minimum molecular weight is required for good foam stability. Similarly, Keuhler and Stine (1974) reported that there is an optimum molecular size for the proteins to be involved effectively in emulsification, and that this specific size was found lower than that of casein. Mita et al. (1977), studying the surface activity of various wheat proteins, related the exceptional foam stability of gluten to its distinguished (high) compressibility, reasonable viscosity and other satisfactory mechanical properties. Dickinson and Stainsby (1982) reported that long-term emulsion stability was usually favored by a mechanically strong film of adsorbed protein around the emulsion droplets.

Davies and Rideal (1961) showed that foam stability was highly related to the surface viscosity of the film as a consequence of the decrease in surface area of the protein.

Disulfide bonds are usually referred to as the main structural barriers against the free change of conformation in proteins (Kinsella, 1981), so their cleavage is expected to allow free movement of the protein subunits.

Richert et al. (1979) correlated the disulfide bond content of proteins to their effect in preventing the higher pH optimum for foam formation, indicating the involvement of disulfide bond interchange in this process. A decrease in bubble size and foam stability of glutenin and other wheat proteins were obtained upon reductive cyanoethylation (Mita et al., 1977). This treatment was accompanied by a significant depression in the surface tension of glutenin; however, no such effect was evident for the gliadin or the gluten fractions.

Undoubtedly, none of the commercially available food proteins has been exclusively adequate as a universal surface active agent. Thus, the need to modify certain properties of these proteins seems indispensable. Enzymatic hydrolysis has been considered one of the most convenient methods for improving the functional properties of food proteins without sacrificing their nutritional value. This process, if carried out under adequate conditions to preserve solubility, control denaturation and enhance hydrophobicity while limiting the amount of polypeptide fragmentation, is expected to result in enhanced molecular flexibility with a minimum amount of damage to their rheological properties.

Gunther et al. (1979) noted that protein derivatives produced by the action of proteolytic enzymes, exhibited adequate aerating and whipping properties. Proteolytic hydrolysis of soybean flour was found to result in a medium-thick, egg white like structure; however, stability of the foam was found to be low (Bobalik and Taranto, 1981). Limited pepsin hydrolysis of soybean protein isolates was found to enhance their foaming and emulsifying capabilities. This effect was found to decrease significantly after a long time of enzyme incubation (Adler-Nissen, 1979; Puski, 1975).

In light of the above discussion, this study was aimed at investigating the effect of a mild enzyme-hydrolytic process on the foaming and emulsifying properties of whey protein isolates (WPI). Enzymes with different specificities were used, namely, trypsin, chymotrypsin and pepsin. Correlation between structural changes in the hydrolyzed WPI molecule and its functional performance in two-related systems, foams and emulsions, was established.

## Materials and Methods:

### 1. Preparation of Whey Protein Isolate Hydrolysates:

Suspensions of 1% WPI (Biopro, Le Sueur, St. Paul, MN) were prepared and adjusted to appropriate pH values prior to enzyme addition (pH 2.0 for pepsin and 8.0 for trypsin and chymotrypsin treatments). Enzymes (Sigma Chemical Co.) were added in the ratio of 0.5:100 (w/w, enzyme to substrate). The hydrolysis process was allowed to take place at 20°C. After 5 hours of incubation,the reaction was terminated by means of heat treatment (90°C for 10 min) and pH adjustment (8.0 for peptic hydrolysates and 3.0 for tryptic and chymotryptic hydrolysates). The resulting hydrolysates were cooled to room temperature and neutralized to pH 7.0. They were further recovered by means of a hollow fiber membrane (AgTechnology Corp., Needham, MA) of MWCO 10,000 followed by freeze drying.

### 2. Conformational and Structural Changes:

Effect of enzymatic treatment on induced structural conformations and flexibility of the WPI molecule was studied in terms of fluorescence shift, hydrophobicity, electrophoretic pattern, surface tension and disulfide bond content.

Fluorescence spectra shifts were used as a measure of the degree of denaturation of the protein as indicated by the change in the number of tryptophan residues in contact with the polar solvent. Native and hydrolyzed WPI dispersions (0.02%) were prepared in pH 8.0 phosphate buffer. Samples were excited at 295 nm and their emission at 400 nm was recorded using a I.S.S. Greg PC photon counting spectrofluorimeter and 1 cm cell pathlength.

Hydrophobicity, expressed as partition hydrophobicity, was determined according to Keshavarez and Nakai (1979). This method was chosen due to the good correlation it bears with functionality. Protein samples were partitioned between two phases dextran (Sigma Chemical Co.) and carbowax or its palmitate ester (Pharmacia,

Piscataway, N.J.). By attaching the palmitate group, carbowax was made more apolar, so the more hydrophobic proteins would show greater tendency for it. In addition, dextran, being a neutral linear polymer would have no complications for polysaccharide gelation or charge-charge interactions with the protein. The concentration of protein in both phases was measured spectrophotometrically at 280 nm against a solution of the corresponding phase without the protein. Partition coefficients of native and hydrolyzed WPI were calculated as:

$$K = C_u/C_l$$

where $C_u$ and $C_l$ represent the concentrations of the protein in the upper and lower phases, respectively. Index of hydrophobicity (DlogK) was calculated as follows:

$$\text{Dlog K} = \log K_1 - \log K_2$$

where $K_1$ and $K_2$ are the partition coefficients in carbowax with and without the palmitate group attached, respectively.

SDS-PAGE was performed in 10% acrylamide, under denaturing conditions (Lammeli, 1970), using a vertical slab gel unit (Hoefer Scientific Inst., San Francisco, CA). Electrophoresis was carried out at 25 mA for 4.5 hours. Proteins were dispersed in distilled water (1%) and 10 microliter samples were injected into the gel slots. Commassie Brilliant Blue R-250 (Bio-Rad, Richmond, CA) in methanol, water and acetic acid was used for staining. Destaining was carried out in two steps using methanol, water and acetic acid. The gels were dried to cellophane films and scanned spectrophotometrically at 600 nm.

Surface tension was measured by means of a Wilhelmy plate balance, which was provided by attaching a thin platinum (Pt) plate to an analytical balance. Protein dispersions (0.5% in pH 8.0 phosphate buffer) were prepared. Measurements were done by monitoring the change in weight of the platinum plate due to the amount of adhering protein solution. Surface tension was calculated as the product of gravitational acceleration and the weight of sample adhered to the platinum plate divided by the perimeter of the plate at the point of contact.

Reduction and blocking of the disulfide bonds were achieved according to Rothenbeuhler and Kinsella (1986). WPI (50 mg) was dispersed in 50 ml nitrogen-saturated tris-HCl (0.03 M. pH 8.0) in the presence of 8 M urea, 1mM EDTA and 0.02% $NaN_3$. Dithiothreitol was added (0.1, 1.0 , 10 and 20 mM) to reduce the disulfide bonds. The reaction mixture was saturated with nitrogen and incubated at 37°C for 6 hours, followed by dialysis (48 hours at 5°C against distilled water) and freeze drying.

Blocking of the reduced disulfide bonds was achieved by dispersing the protein (50 mg) in 50 ml (0.03 M) tris-HCl buffer followed by addition of 0.23 mM sodium-tetrathionate. After 1 hour of incubation, the reaction mixture was dialyzed against distilled water and freeze dried. Total disulfide bond content of native, hydrolyzed and s-sulfenyl modified WPI was determined according to the method of De Marco et al. (1966). Complete breakdown of the disulfide bonds was done by cyanolysis (KCN) in 1 N ammonia, followed by coupling with N-ethylmaleinamide. Absorption at 300 nm was recorded using cysteine-HCl as a reference.

Viscosity of 5% protein dispersions was measured using a Haake Rotovisco rotational viscometer (Haake Inc., Saddlebrook, N.J.) Model RV-3, using an NV sensor. All the measurements were carried out at 25°C.

3. Foaming Ability and Stability:

Protein dispersions (0.5% in distilled water) were prepared and whipped in a Waring blender for 5 minutes at ca. 17,000 rpm. Foam expansion was expressed as the ratio of the volume of the fluid after whipping to its initial volume, multiplied by 100. Foam stability was determined similarly for whipped samples after setting undisturbed in the same graduated cylinder for 30 minutes.

4. Emulsifying Properties:

Emulsifying capacity was determined in terms of oil droplet size distribution using a Coulter counter (Coulter Scientific Inst., Hialeah, FL) Model ZBI, equipped with a 100 $\mu$m orifice. A Sodium chloride solution (0.18 M) was used as a suspending electrolyte and latex suspensions (Duke Scientific Corp., Palo Alto, CA) of 1, 4 and 7 $\mu$m diameter as standards. The ultrastructure of the emulsions prepared from native or hydrolyzed WPI was observed using cryo-scanning electron microscopy (Amray 1000A). Droplets of freshly prepared emulsions 20:80 corn oil to protein (1% in water) were placed in wells of copper specimen stubs and quench-frozen by dipping into a nitrogen-chilled propane (approximately -185°C). Fracturing was done using a cold-edge knife and allowed to etch for an appropriate time. The surface of the sample was then coated with gold/palladium (40/60) and examined under the microscope.

RESULTS and DISCUSSION

1. Conformational Changes:

Fluorescence spectra of native and hydrolyzed WPI are presented in figure 1. Native WPI showed a fluorescence peak at ca. 330 nm, typical of a globular inflexible protein (Freifelder, 1982). All the hydrolytic treatments resulted in enhanced peak shift towards longer wavelength, indicating the exposure of more tryptophan residues to the outer polar phase of the medium. Among the samples tested, chymotryptic hydrolysates showed the greatest red shift followed by the tryptic and peptic ones. Estimation of the degree of denaturation achieved by the enzymatic treatments was done by running a comparative spectrum for native WPI in 8 M urea (well-known protein denaturant and hydrophobic interaction perturbant). The fluorescence peak of chymotrypsin-treated WPI was found to be fairly superimposable on that of the urea-treated WPI which implies that chymotrypsin had denatured the protein most effectively.

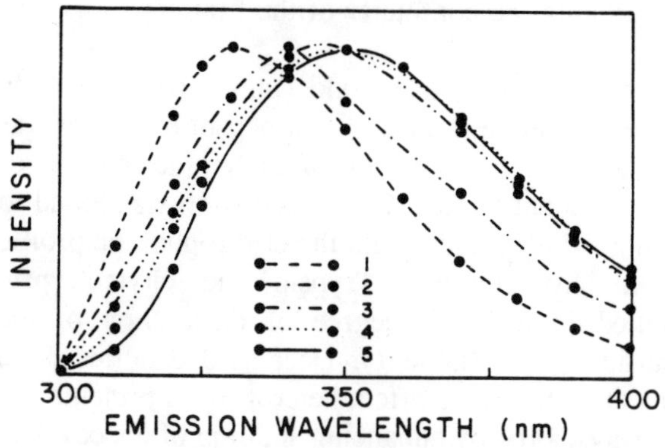


Figure 1. Fluorescence spectra of native (1), pepsin-treated (2), trypsin-treated (3), chymotrypsin-treated (4) and urea-treated WPI (5). Protein (0.05%) in phosphate buffer, pH 8.0.

Surface hydrophobicity determination, expressed as "partition hydrophobicity", shows that peptic WPI hydrolysates resulted in the highest hydrophobicity coefficient compared to all of the proteins tested (figure 2). The hydrophobicity coefficient of native WPI was found to be the lowest. This result supports the common understanding that configurational stability of these proteins in their native state (globular) is not simply controlled by the distribution of polar to apolar residues on both sides of the molecule and the clustering of charged groups on the surface (since thsi procedure eliminates charge differences), but rather due to the energetically favorable arrangement of residues within the core of their structure i.e. H-bonding and hydrophobic interactions (Dickinson, 1986).

Pepsin, due to its high specificity to cleave peptide bonds adjacent to hydrophobic amino acid residues, is expected to result in a significant disengagement of these fluorophores (tryptophan mainly) from their chains and thus, facilitates their orientation into the apolar site of the medium. In addition, the highly acidic conditions required for the activity of pepsin (pH 2.0) are expected to reduce the majority of the covalent bonds, thus acting synergistically to expose more hydrophobic groups from within the enclosed globule. Trypsin and chymotrypsin hydrolysates, although showing a considerable increase in hydrophobicity coefficient compared to the native protein, were inferior to pepsin. Trypsin and chymotrypsin, in addition to their hydrolytic activity are powerful deamidating proteases (Kato et al., 1987) and thus are efficient in breaking the joint between free amino and neighboring carboxyl groups. This appears to be in good agreement with their fluorescence behavior.

The effect of enzyme hydrolysis on the molecular weight changes in WPI is shown in figure 3. The scanned electrophoretic profile shows, in general, that all the hydrolytic treatments resulted in lower molecular weight fractions compared to the native WPI. This feature was most demonstrated by pepsin, followed by trypsin and chymotrypsin. Electrophoresis patterns show that all the enzymes employed in this study have high affinity towards the high molecular weight subunits of the protein. This is reflected in smaller areas of the peaks corresponding to the immunoglobulins and bovine serum albumin, in particular. Pepsin-treated samples showed a significant decrease in the β-lactoglobulin fraction. This may be due to two factors. First, the enzymes' inherent affinity for this fraction and the highly acidic medium of the hydrolysate preparation, which aids in the destruction of the β-lactoglobulin subunit (McKenzie, 1971)

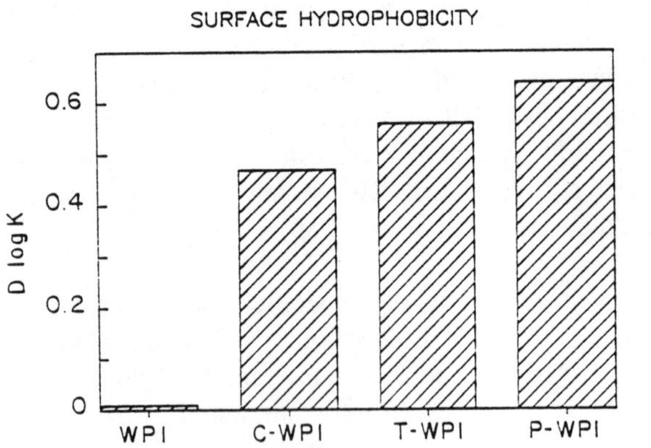


Figure 2. Partition hydrophobicity coefficient (Dlog K) of native (WPI), chymotrypsin-treated (CWPI), trypsin-treated (TWPI) and pepsin-treated (TWPI).

Surface tension measurements as a function of pH are shown in figure 4. Native WPI resulted in the highest surface tension along the whole pH range tested, while peptic ones were the lowest. This reflects a higher molecular flexibility (Fry and Poole, 1985). The high content of the low molecular weight peptides in peptic hydrolysates, added to their enhanced hydrophobicity, might have facilitated their diffusion into the interfacial layer. Furusawa et al. (1982) reported on the high capability of the low molecular weight peptides to cross the interfacial film between the aqueous and the hydrophobic layers, instantly, though with time, these small peptides were found to be replaced by higher molecular weight components.

Griffin et al. (1984) showed that upon fragmentation of kappa-casein, by means of disulfide bond reduction with 2-mercaptoethanol, their penetration into a lipid monolayer was enhanced significantly. It is believed that for food proteins, in general, the most important driving force for surface adsorption is caused by hydrophobic destabilization of their molecular structure.

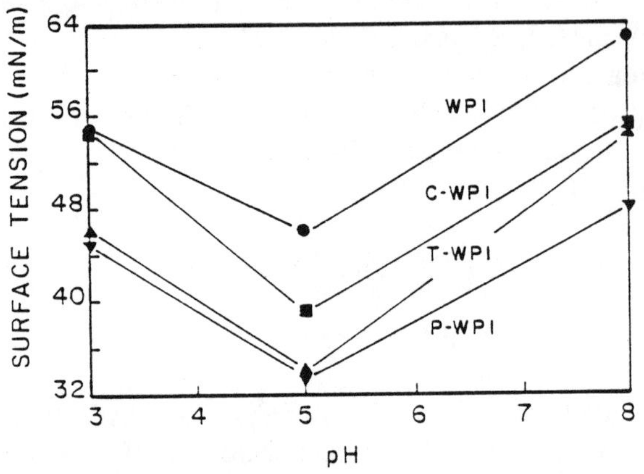


Figure 4. Surface tension at air-water interface of native (WPI), chymotrypsin-treated (CWPI), trypsin-treated (TWPI), and pepsin-treated (PWPI) whey protein isolates.

The effect of pH on changes in surface tension was found to follow the same trend for all the protein samples, i.e. a minimum around the isoelectric point and a significant increase on both sides of the profile away from pH 5.0. At the isoelectric point, proteins usually self-associate most strongly, so steric or entropic forces of repulsion are minimum (Bull, 1948). In addition, maximum intermolecular attraction leads to maximum activation energy and eventually to enhanced surface activity by means of encouraging the continuity of the film.

Changes in disulfide bond content caused by the hydrolytic process showed that pepsin-treatment was the most drastic in terms of higher destruction of these covalent bonds. This observation was found to be in good agreement with the electrophoretic profile of the hydrolysates. Trypsin and chymotrypsin resulted in a lower degree of destruction of the disulfide bonds (table 1). Decreased disulfide bond content is known to affect heterogenous proteins in two ways. First by enhancing splitting of the covalent bonds, thus lowering the chances for protein-protein interaction. Secondly, by decreasing viscosity as a direct consequence of the subunits' disengagement.

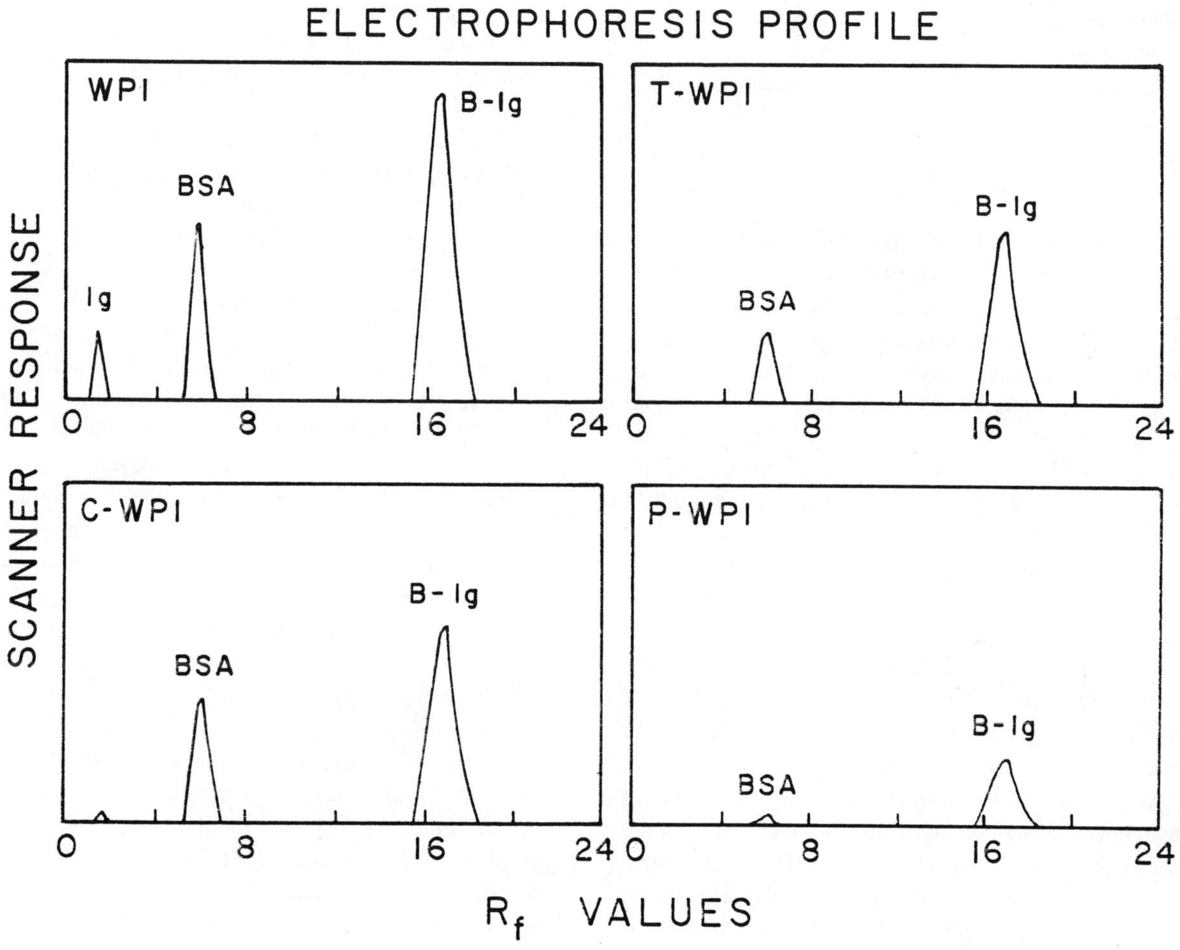


Figure 3. Scan of electrophoretograms of native (WPI), chymotrypsin-treated (CWPI), trypsin-treated (TWPI) and pepsin-treated (PWPI) whey protein isolates at 600 nm.

Table 1. Disulfide Bond Content (%) of Native and Hydrolyzed WPI.

| PROTEIN | % -SS |
|---|---|
| Native | 100.0 |
| Chymotrypsin-treated | 85.3 |
| Trypsin-treated | 80.1 |
| Pepsin-treated | 60.5 |

2. Foaming Properties:

Foaming properties of native and hydrolyzed WPI are shown in figure 5. Hydrolytic treatment is shown to enhance the foaming ability of WPI significantly. Peptic hydrolysates exhibited the highest degree of foam expansion (as measured instantaneously after whipping), while trypsin and chymotrypsin showed an inferior power. For peptic hydrolysates, the prevalence of low molecular weight fractions and the enhanced accessibility of hydrophobic residues, might have facilitated the protein unfolding and increased its capability of entrapping more air bubbles. With trypsin and chymotrypsin treatments, the increase in hydrophobicity and molecular weight reduction were more moderate. This seems to be fairly in accordance with their foaming behavior. Although surface hydrophobicity is normally considered an important factor in the onset of foaming, many workers suggested a differentiation between random hydrophobicity and surface hydrophobicity due to the close association of the latter to functional properties of proteins (Keshavarez and Nakai, 1976). Poole and Fry (1987) noted that surface hydrophobicity, if accompanied by preserved solubility, is very likely to result in a highly functional protein. Poole et al. (1984) showed that surface hydrophobicity of ovalbumin adjusted to pH 7.4 after heating at 80°C for 10 minutes was increased significantly (from 20 to 2200) as determined by the cis-parinaric acid fluorescence probe technique. However, foaming properties were found to be inferior to those of the native (surface hydrophobicity 1450), indicating that other structural features were involved in the foaming process.

Several reports on the positive effects of hydrolytic treatment on the foaming ability of different food proteins have confirmed the importance of these basic requirements (Bobalik and Taranto; Eldridge, 1963; Jost and Monti, 1982). However, a new insight into the effect of some particular peptide terminals on the overall surface activity of the protein molecule was offered by Shimuzi et al. (1983). Their study showed that functionality of casein is highly dictated by the hydrophobicity of a specific peptide i.e. the N-terminal of the kappa-casein fraction.

Data on foam stability showed that pepsin-hydrolysates exhibited the lowest stability while those of chymotrypsin were the highest, i.e. the reverse order to foam expansion. Peptic hydrolysates, in addition, were even less stable than the native proteins. The theory of polymer adsorption clearly predicts that for a polydisperse polymer, adsorption of shorter chains is kinetically favorable, which explains simply, the superiority in foam expansion of the peptic hydrolysates (low molecular weight) over the rest of the proteins tested. However, with time, polymers of larger molecular weight are thermodynamically driven to replace the smaller ones and thus provide film stability (Cohen et al., 1980; Koopel, 1981). Statistically, this effect arises from the differential balance between energy and entropy contributions for short- and long-chain polymers (Dickinson, 1985).

Viscosity associated with higher molecular weight peptides, might have also accounted for stable film formation (table 2). This behavior is, in fact, similar to that of gums and other food stabilizers whose effect on enhanced foaming and emulsifying properties of food systems is based on their characteristic stabilization of the polymer lattices, but not on surface activity, since these compounds do not have any significant impact on surface tension (Bergenstall, 1986). For food proteins, in general, the stronger the tendency towards aggregation might favor the domination of larger fragments as contributors to film stability. This effect has been experienced with many food proteins, especially with casein micelles and their component fragments (Dickinson, 1982).

Table 2. Viscosity of Native and Hydrolyzed WPI (5%) in distilled water.

| PROTEIN | VISCOSITY (cP) |
|---|---|
| Native | 3.35 |
| Chymotrypsin-treated | 3.41 |
| Trypsin-treated | 3.24 |
| Pepsin-treated | 2.87 |

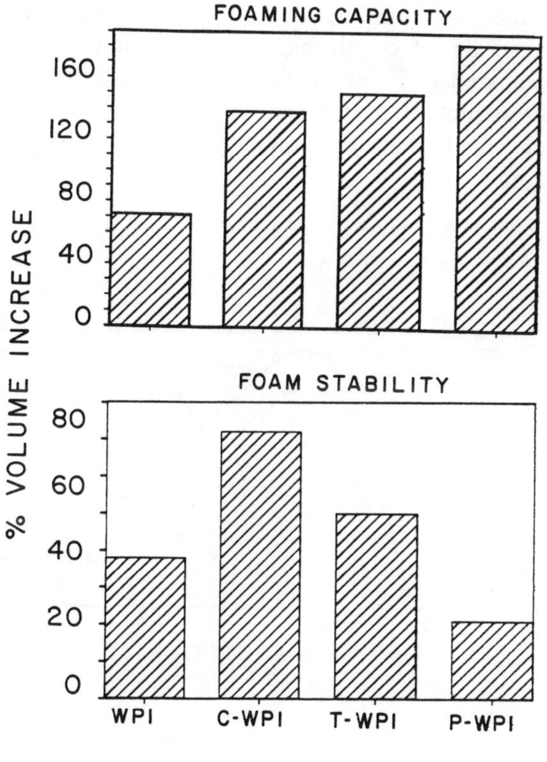


Figure 5. Foaming capacity and stability of native (WPI), trypsin-treated (TWPI), chymotrypsin-treated (CWPI) and pepsin-treated (PWPI) whey protein isolates (0.5%) in distilled water.

Figure 6 compares the effect of disulfide bond reduction on the overall foaming capacity and stability of foams. Blocking the disulfide bonds of WPI and the corresponding production of s-sulfenyl residues clearly show a significant increase in foam expansion. This confirms the idea that low molecular weight components of the protein are useful in initiating the foaming process, while their high molecular weight counterparts are important for maintaining the foam structure. Disulfide bonds provide this activity by promoting strong coherent protein layers and therefore, stable foams (Mita et al., 1978). The importance of induced disulfide-linked complex of $\beta$-lactoglobulin and kappa-casein on the baking properties of milk powders has been demonstrated (Halling, 1981).

Addition of 5% native WPI to the hydrolysates showed an enhanced foaming stability which dictates the importance of high molecular weight components for film cohesiveness (figure 7). Kinsella (1979) demonstrated that soy globulins, because of their molecular size, diffuse relatively slowly, but once at the interface, they spread easily through subsequent penetration into the film. Another point that should be taken into consideration is that, addition of 5% protein also involves the addition of more solids to the mixture which accounts for higher film stability.

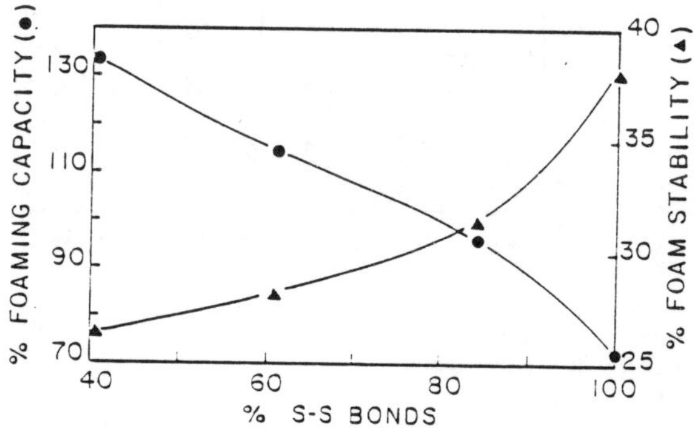


Figure 6. Effect of disulfide bond content on the foaming ability and stability of WPI.

The combined effect of protein concentration and pH on foaming capacity of native and hydrolyzed WPI is presented in figure 8 (a-d). The general feature of this experiment shows that for all the preparations tested, foaming capacity increased progressively with the increase in protein concentration. For native and chymotryptic WPI, a slight decline in foam expansion was observed at higher concentration. The prevalence of higher molecular weight components in these proteins might have led to detrimental denaturation upon whipping and eventually resulted in protein aggregation. Similar observations have been reported in the literature (Halling, 1981; Kinsella, 1979; Richert, 1979), though no upper limit of concentration can be drawn from data on different proteins due to the interference of other factors in the foaming process.

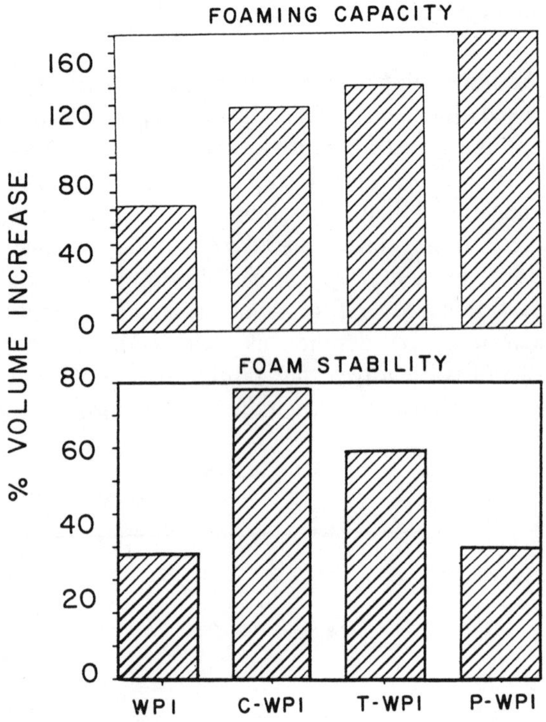


Figure 7. Effect of addition of 5% native WPI on the change in foaming ability and stability.

Foaming-pH profiles show, for all proteins examined, that maximum foam expansion was obtained in the vicinity of the isoelectric point of the protein (pH 5.0). This observation falls in good agreement with the data on surface tension. The effect of pH on proteins is usually explained based on two main factors; these are net charge of the system and protein conformation. As explained earlier, at the isoelectric point, there is minimum charge-charge repulsion and maximum protein-protein interaction which might have affected positively the formation of a continuous film (Graham and Phillips, 1976; Waniska and Kinsella, 1979). An interesting observation in these data is that peptic hydrolysates possessed the highest foam expansion along the whole pH range studied. This might imply that surface activity is the most dominant factor in determining the foaming behavior of a given protein and that this aspect might counteract others resulting from protein-protein repulsion at pH values away from the isoelectric point.

3.Emulsifying Capacity:

Droplet size distribution of oil in emulsions prepared from native and hydrolyzed WPI is shown in figure 9. All the treatments resulted in a significant tendency towards smaller oil droplets. Native WPI showed an uneven, skewed profile with a major clustering of oil droplets in the vicinity of 4 $\mu$m compared to 2 $\mu$m for the hydrolysate preparations. Ultrastructure of the micrographs shows a similar behavior to that of the oil droplet distribution profile (figure 10).

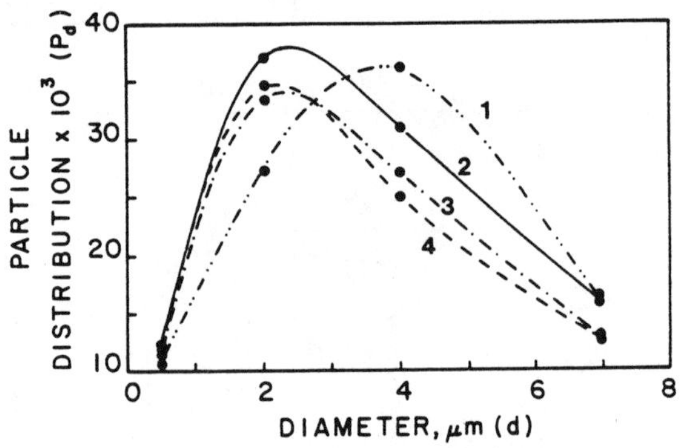


Figure 9. Droplet size distribution of oil in emulsions prepared from native (1), chymotrypsin-treated (2), trypsin-treated (3) and pepsin-treated (4) whey protein isolates. (Pd, is the number frequency and d$\mu$, droplet diameter)

Thickness of the film varied widely among the different preparations investigated. Emulsions prepared from native WPI resulted in a very wide distribution of oil droplets within a plaque-like protein lining. This behavior confirms the lack of flexibility predicted for these globular proteins from their high surface tension and low degree of denaturation (lower

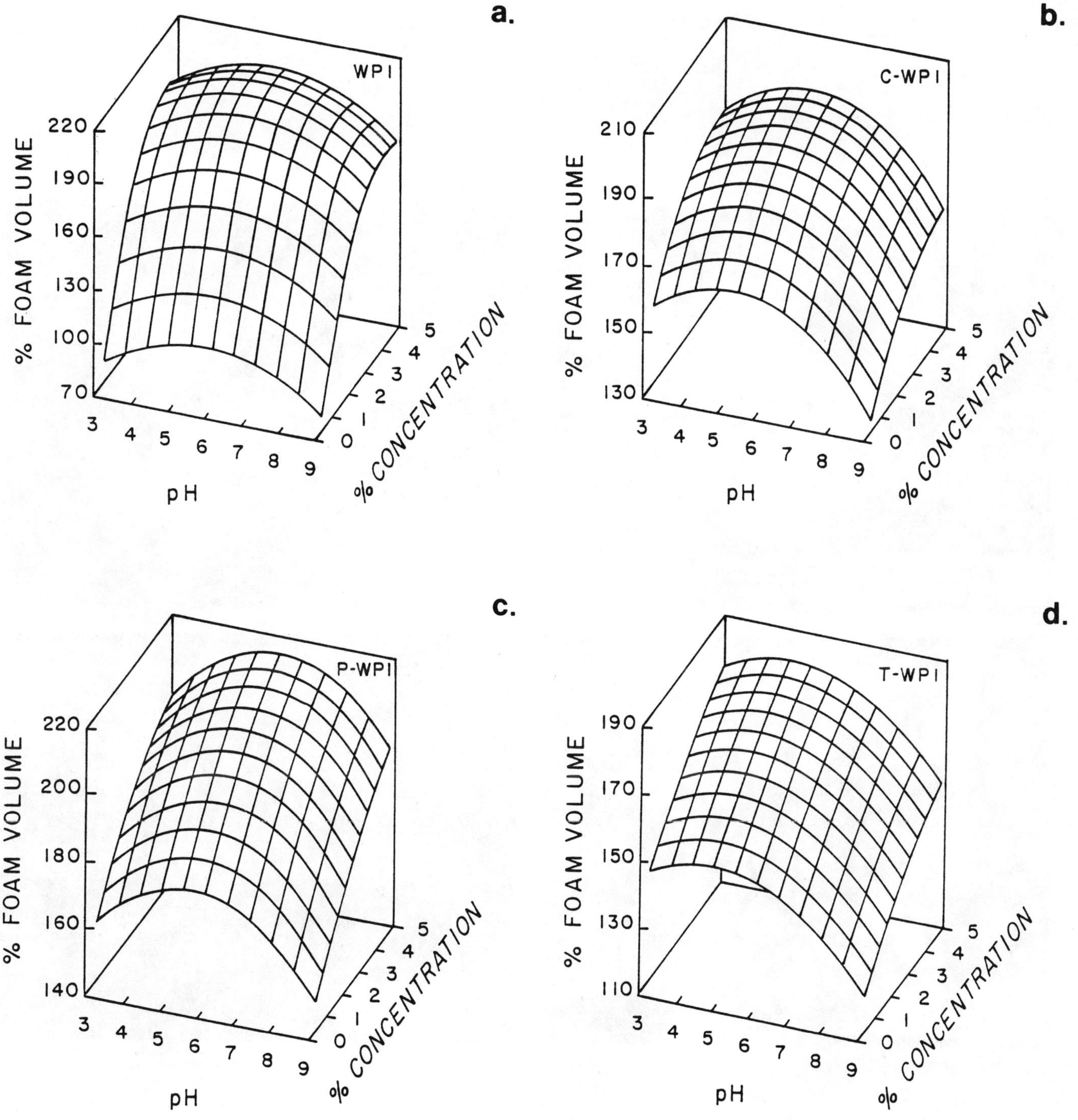


Figure 8. Response Surface Methodology plots for the effect of protein concentration and pH on the foaming ability of native (a), chymotrypsin-treated (b); pepsin-treated (c) and trypsin-treated (d) whey protein isolates.

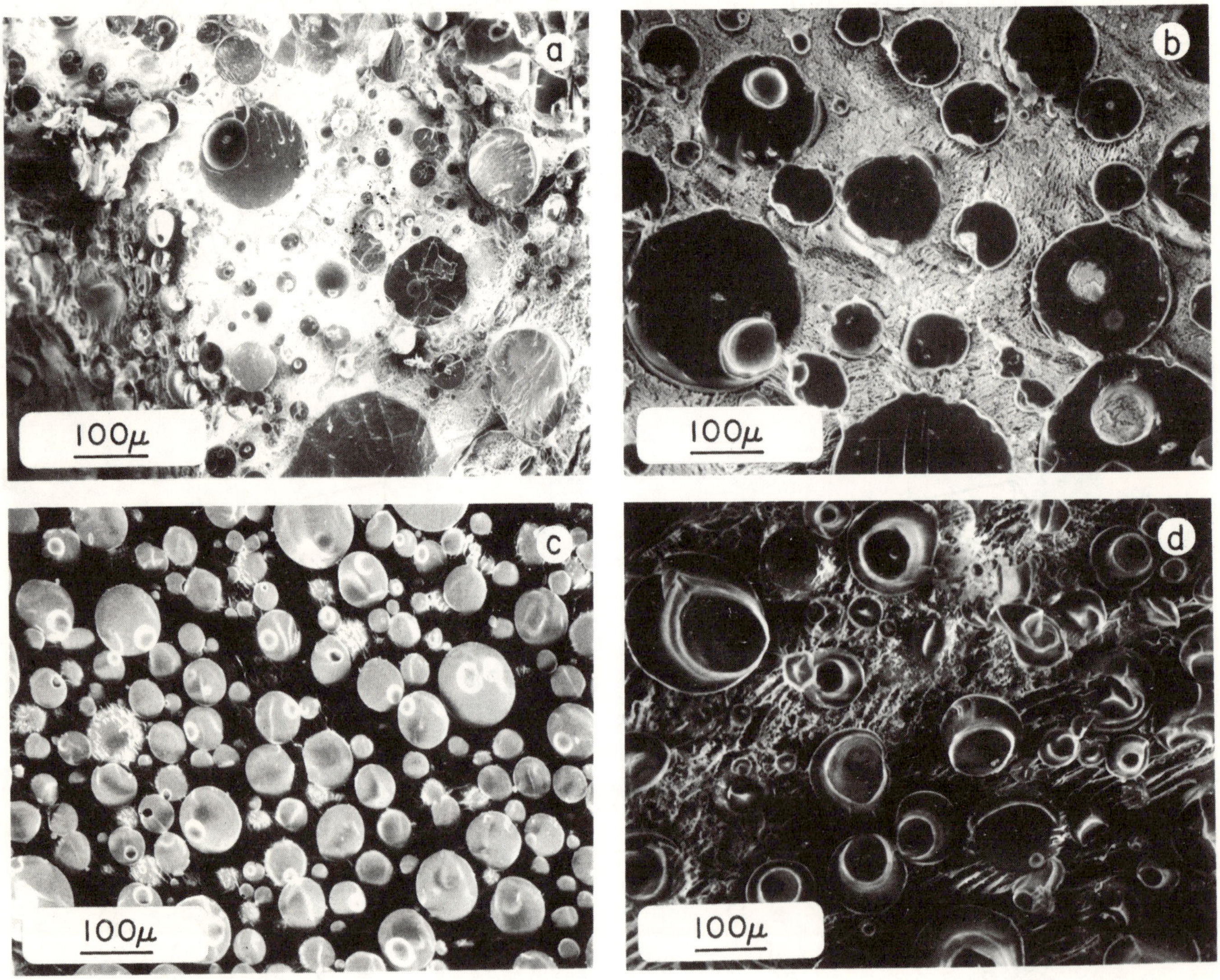


Figure 10. Scanning Electron Micrographs for emulsions of corn oil in native (WPI), chymotrypsin-treated (TWPI), trypsin-treated (TWPI) and pepsin-treated (PWPI) whey protein isolates (oil:protein 20:80 w/w).

fluorescence wavelength). The micrograph of chymotryptic hydrolysates showed a fair distribution of round oil droplets within the protein film. Two factors might be responsible for this desirable behavior: first, the moderate decrease in molecular weight and the formation of highly denatured peptides might have allowed a better association of the hydrophobic sites in the protein film to the oil droplets. Such a film should have enough flexibility to resist osmotic pressure and so avoids droplet bursting. Trypsin- and pepsin-hydrolysates showed a similar droplet size distribution behavior as measured by oil droplet counting. However, a closer look at the micrographs revealed the possibility of two different events for both treatments. For pepsin, the film appears to be thin (diluted) while that of trypsin is rougher and inflexible. Both cases are expected to lead to emulsion destabilization. This feature represents a good example of almost equivalent number of oil droplets, but different ultrastructure of the emulsion. This observation stresses the importance of the protein inherent characteristics on its emulsifying performance. Data from different sources on the capability of hydrolyzed proteins to enhance emulsification vary widely. In some cases, hydrolysates were found to outperform their native proteins in emulsion formation (Adler-Nissen, 1976; 1972; Toguichi et al., 1982; Zakaria and McFeeters, 1977), while in others, this treatment was found to adversely affect the emulsifying behavior of the protein (Halling, 1981; Keuhler and Stine, 1974; Puski, 1975). It appears that an optimum degree of hydrolysis is required for improved emulsification properties which varies with the type of protein and other inherent characteristics such as hydrophobicity, denaturation or others.

ACKNOWLEDGEMENT:

The experiments and analyses of the fluorescence data were performed at the Laboratory for Fluorescence Dynamics (LFD) at the University of Illinois at Urbana-Champaign (UIUC). The LFD is supported jointly by the Division of Research Resources of the National Institute of Health (RR03155-01) and UIUC). Our thanks are extended to Ms. Ruth Rusch for the technical assistance in the fluorescence experiment.

REFERENCES

Adler-Nissen, J., Enzymatic hydrolysis of proteins for increased solubility, J. Agric. Food Chem. 24 (6), 1090, (1976).

Bergenstall, B., Folgeri, S. and Stenin, P., in Gums and Emulsion Stabilizers for the Food Industry., v.3, p. 285, Phillips, G.O., Wedlock, D. and Williamson, P.,(Eds.) Elsevier Appl. Sci., London, (1986).

Bobalik, J. and Taranto, M., The effect of enzymatic modification on the foaming, water absorption and baking qualities of defatted soya flour. J. Food Technol. 15, 637 (1980).

Bull, M., Spread monolayers of protein. Adv. Prot. Chem., vol III, p.95, Acad. Press, New York, (1948).

Cumper, C., The stabilization of foams by proteins. Trans. Faraday Soc. 49, 1360 (1953).

Davies, J. and Rideal, E., in Interfacial Phenomena, p.366. Acad. press, New York, (1961).

De Marco, C., Graziani, M. and Mosti, R., Spectrophotometric determination of thiols and disulfides with N-ethylmaleinimide. Anal. Biochem. 15, 40 (1966)

De Vilbiss, E., Holsinger, V., Posati, L. and Pallansch, M., Whey protein concentrate foams. Food Technol. 28 (3), 40 (1974).

Dickinson, E., Mixed proteinaceous emulsifiers: review of competitive protein adsorption and the relationship to food colloid stabilization. Food Hydrocolloids. 1 (1), 3 (1986)

Dickinson, E. and Stainsby, G. in Colloids in Foods, 533 pp. Appl. Sci. Publ., London (1982).

Fox, P., Morrissey, P. and Mulvhill, D., in Developments in Food Proteins, vol. 1, p.1, Appl. Sci., London, (1982).

Frefelder, D., in Physical biochemistry, p. 537, second edition, Freeman, W. and Co. (Ed.), San Francisco, (1982).

Furusawa, K., Yamashita, K. and Konno, K., Adsorption of monodisperse polystyrene onto glass. I. Preference adsorption and displacement of high molecular weight species. J. Colloid Interface Sci. 86, 35 (1982).

Graham, D. and Phillips, M., Proteins at liquid interfaces.I. Kinetics of adsorption and surface denaturation. J. Colloid Interf. Sci. 70, 403 (1979).

Graham, D. and Phillips, M., in Foams, p. 237, Akers, C. (Ed.), Acad. Press, London (1976).

Gunther, R., Chemistry and characteristics of enzyme-modified whipping proteins. J. Am. Oil Chem. Soc. 56, 345 (1979).

Halling, P., Protein-stabilized foams and emulsions. C.R.C. Rev. Food Sci. Nutr. 15, 153 (1981).

Horiuchi, H., Fukushima, D., Sugimoto, H. and Hitori, T., Studies on enzyme-modified proteins as foaming agents: Effect of structure on foam stability. Food Chem. 3, 35 (1978).

Johnson, T. and Zabik, M., Egg albumen proteins interactions in an angel food cake system. J. Food Sci. 46, 1231 (1981).

Jost,R. and Monti,J., Emulgateurs peptidiques obtenus par l'hydrolyse enzymatique partielle de la proteine serique du lait. Lait 62, 521 (1982).

Kato, A., Tanaka, A., Matsudoni, N. and Kobayashi, K., Deamidation of food proteins by proteases in alkaline pH. J. Agric. Food Chem. 35, 224 (1987).

Kato, A., Matsudo, T., Matsudoni, N. and Kobayashi, S., Determination of protein hydrophobicity using a sodium dodecyl binding method. J. Agric. Food Chem. 32, 284 (1984).

Kato, A. and Nakai, S., Hydrophobicity determined by a fluorescence probe method and its correlation with surface properties of proteins. Biochim. Biophys. Acta 624, 13 (1980).

Keshavarez, F. and Nakai, S., The relationship between hydrophobicity and interfacial tension of proteins. Biochim. Biophys. Acta 576, 269 (1979).

Keuhler, C. and Stein, C., Effect of enzymatic hydrolysis on some functional properties of whey protein. J. Food Sci. 39, 379 (1974).

Kinsella, J., Functional properties of proteins: Possible relationship between structure and function in foams. Food Chem. 7, 273 (1981).

Kinsella, J., Functional properties of soy proteins. J.Am. Oil Chem. Soc. 56, 242 (1979).

Koopel, L., The effect of polymer dispersity on the adsorption isotherm. J. Colloid Interface Sci. 83, 116 (1981).

Lammeli, V., Cleavage of structural proteins during the assembly of the head of bacteriophage T4. Nature 227, 680 (1970).

Mansvelt, J., Technical application of protein whipping agents in sugar confectionery. J. Am. Oil Chem. Soc. 56, 350 (1979).

McKenzi, H., in Milk proteins, Chemistry and Molecular Biology, vol. 1 & 2, Acad. Press, New York (1970).

Mita, T., Ishido, E. and Matsumoto, H., Physicochemical studies on wheat protein foams. J. Colloid Interface Sci. 64 (1), 143, (1977).

Mitchell, J., In Developments in Food Proteins, vol 4, p.290. Hudson, B. (Ed.), Elsevier Appl. Sci., London (1986).

Muschiolik, G., Dickinson, E., Murray, B. and Stainsby, G., Interfacial and emulsifying behavior of acetylated field bean protein isolate. Food Hydrocolloids 1 (3), 191 (1987).

Phillips, M., Protein conformation at liquid interfaces and its role in stabilizing emulsions and foams. Food Technol. 35, 50 (1981).

Poole, S., West, S. and Walters, C., Protein-protein interactions: Their importance in the foaming of heterogenous protein systems. J. Sci. Food Agric. 35, 701 (1984).

Puski, G., Modification of functional properties of

soy proteins by proteolytic enzyme treatment. Cereal Chem. 52, 655 (1975).

Richert, S., Physical-chemical properties of whey protein foams. J. Agric. Food Chem. 27 (4), 665 (1979).

Rothenbuhler, E. and Kinsella, J., Disulfide reduction and molecular dissociation improves the proteolysis of soy glycinin by pancreatin. J. Food Sci. 51 (6), 1479 (1986).

Shimizu, M., Lee, S., Kaminogawa, S. and Yamauchi, K., Emulsifying properties of an N-terminal peptide obtained from the peptic hydrolysate of α-s1 casein. J. Food Sci. 49, 1117 (1983).

Toiguchi, S., Maeda, S., Watanabe, M. and Arai, S., Purification and characterization of a surface-active macropeptide produced by modification with papain. Agric. Biol. Chem. 46 (12), 2945, (1982)

Waniska, R. and Kinsella, J., Enzymatic hydrolysis of maltosyl and glucosaminyl derivatives of β-lactoglobulin. J. Agric. Food Chem. 32, 1042 (1984).

Zakaria, F. and McFeeters, R., Improvement of the emulsification properties of soy protein by limited pepsin hydrolysis. Lebensm. Wissen. Technol. 11, 4222 (1978).

# MOBILITY OF WATER IN FOAMS DETERMINED BY MAGNETIC RESONANCE SPECTROSCOPY

M.J. McCarthy and J.R. Heil ■ Department of Food Science and Technology, University of California, Davis, CA 95616

Nuclear magnetic resonance studies of foams showed that the spin-spin relaxation time varies in the foam as a function of density. This variation is described in terms of a classic two-state fast exchange model. Procedure for estimation of localized spin-spin relaxation times and the limitations of both, procedure and model, are discussed.

Nuclei which have a magnetic moment can be polarized in a large static magnetic field. The equilibrium magnetization is proportional to the density of the nuclei being considered. In pulsed Nuclear Magnetic Resonance (NMR) experiments, the magnetic equilibrium is disturbed by a characteristic resonance radio frequency (Larmor frequency). The pulsed energy absorbed, exciting the nuclei, is released as the nuclei relax to their magnetic equilibrium. The process of relaxation is described by two time constants; spin-lattice or longitudinal relaxation ($T_1$), and spin-spin or transverse relaxation ($T_2$) times. In other words, $T_1$ describes the nuclei's return to the ground state, (loss of longitudinal magnetization), and $T_2$ describes dispersion of spins, (loss of transverse magnetization). Both time constants are dependent on correlation time, $\tau_c$, a measure of mobility for the nuclei under consideration (1).

Department of Food Science and Technology, University of California, Davis, California

In magnetic resonance imaging and spectroscopy, Signal intensity (S) is directly proportional to the density ($\rho$) of nuclei being observed. However, signal intensity will also depend on both experimental and relaxation time parameters:

$$S \alpha \rho \,[\exp(1 - {}^{PD}/_{T_1})]\,[\exp(-{}^{TE}/_{T_2})] \qquad (1)$$

Of the experimental parameters, PD is the delay time between acquisitions, and TE is the delay time between the $\pi/2$ and $\pi$ pulses used in the 2D-Fourier transform spin-echo sequence. Relaxation times $T_1$ and $T_2$ are characteristics to the material, and are directly related to the mobility of the observed nuclei within the sample.

An overview by Weisser (2) described the utility of $T_1$ in monitoring freezing/concentrating of food systems, and in characterizing saturated-unsaturated oil and water fractions in edible fats and oil seeds. Lillford et al. (3) used both $T_1$ and $T_2$ time constants to characterize distribution of various states of mobility of water in heterogeneous foods. Richardson and Steinberg (4) presented application of relaxation parameters to measuring binding of water in a number of carbohydrate and proteinaceous systems. These authors also discussed factors which influence experimental results when measuring mobility of water.

The purpose of this work was to devise a Magnetic Resonance Imaging (MRI) procedure for measuring the mobility of water as a function of density during draining of liquid egg-white foam, and to evaluate the results in terms of an existing two-state model.

## EXPERIMENTAL

### Sample Preparation

Dried egg whites (Henningsen Foods, Inc., Omaha, NE) were rehydrated using one part egg whites and seven parts of distilled water. The samples were refrigerated for minimum 12 h before an experiment to assure that rehydration was completed. For each test, 50 mL. of the reconstituted egg whites were beaten two min at high speed, with a five-speed hand mixer. The sample holder was filled with the whipped foam by scooping its open ended cylinder into the foam until filled. The cylinder was leveled and the outside wiped of adhering foam. The bottom of the cell was attached, and the sample was placed into the magnet. The five cm diameter by five cm high cell had a plastic screen with two $mm^2$ openings placed onto the bottom of the cylinder, providing free drainage. The separated and fallen liquid was captured in a secondary cell attached to the foam sample cylinder. Sample holder and a representative one dimensional image are schematically illustrated in Figure 1.

### Magnetic Resonance Imaging

A General Electric two-Tesla chemical shift multinuclear magnetic resonance spectrometer and imaging system was used to obtain proton density projections at 85.5 MHz resonance frequency. Field homogeneity, the 180° pulse length and pulse power were determined on liquid egg whites. Echo delay times of 10 and 20 ms were alternated for each set of acquisitions in a 2D-Fourier transform-spin echo sequence with the phase encode gradient turned off. Predelay time of ten sec and minimum four acquisitions were used. Data were acquired at two min. intervals, for a maximum period of 96 min. Times of spin-lattice relaxation ($T_1$) and spin-spin relaxation ($T_2$) were determined using the $\pi - \tau - \pi/2$ inversion-recovery and the $\pi/2 - \tau - \pi - \tau$ sequence, respectively (5), after various periods of drainage.

## RESULTS AND DISCUSSION

Spin lattice relaxation times determined by the inversion-recovery sequence were repeatedly characterized by a single value ($T_1 = 1.65 \pm 0.02$ s) in both the liquid egg whites and the draining foams. Therefore, it may be concluded that the $T_1$ values in draining egg white foam, being the same in both the liquid and the foam, did not vary as a function of density. Thus, based on the relationship shown in Equation 1, Predelay Time (PD) was set at over $5 \star T_1$ to reduce the influence of longitudinal relaxation on signal intensity to a negligible level in subsequent experiments.

Typical spin-spin relaxation ($T_2$) curves obtained on liquid egg whites and the corresponding foam are shown in Figure 2.
The data obtained for the liquid could be characterized by a single value. However, the data for the foam changed with time of draining, and required fitting with at least two exponential curves, indicating the existence of multiple phases, i.e., different states of water proton mobility. Since during a draining period, distribution of density continually changed in the sample, the multiple exponential fit

to the spin-spin relaxation data implied that the apparent water proton mobility in the foam varied as a function of density. Therefore, it was expected that $T_2$ would change in both time and position in the draining sample. As transverse relaxation couldn't be characterized by a single value, the $T_2$ correction of the MRI signal intensity needed further consideration.

From the relationship in Equation 1 where water proton density was considered constant and PD=10 s made the expression for the $T_1$ correction approach unity, an equation for calculating positional $T_2$ values was derived:

$$T_2 = \frac{(TE_1 - TE_2)}{\log(\frac{S_2}{S_1})} \qquad (2)$$

$T_2'$ is an estimated positional spin-spin relaxation time; $TE_1$ and $TE_2$ are two different echo delay times; and $S_1$ and $S_2$ are positionally corresponding points on the respective 1-D images obtained with $TE_1$ and $TE_2$. Examples of one dimensional images obtained with 10 and 20 s echo delay times are shown in Figure 3. Taking spatially corresponding signal intensities from each projection, positional $T_2'$ values were calculated. Using these $T_2'$ values, signal intensities were corrected as follows:

$$\text{corrected signal intensity} = \frac{S_1}{\exp(\frac{-TE_1}{T_2'})} \qquad (3)$$

In an isolated system, such as the draining foam in this experiment, the distribution of spins is expected to change, but their sum should remain constant, unless significant changes develop in the mobility of the spin active nuclei. The soundness of the procedure for estimating positional $T_2$ and the need for correcting for spin-spin relaxation are both demonstrated in Figure 4. For a period of drainage where no liquid separation was observed, the sums of uncorrected signals from the foam were not only significantly smaller in magnitude than the $T_2'$ corrected signals, but they continually increased with time. The observed increase is another good indication that water mobility was not constant in the foam system. In comparison, the sums of the corresponding $T_2'$ corrected signals have remained relatively constant, showing that positional estimations of $T_2$ in this work were reasonably consistent with the actual trends in spin-spin relaxation, i.e., in water mobility.

Egg white foam is assumed to be a wet foam with polyhedral bubbles. The bubbles are separated by films between the faces, and Plateau borders at the intersection of the faces (6). Draining of the foam involves the liquid flowing along the films into the borders, following the direction of gravity. As the liquid in the film decreases, the film reaches a critical thickness at which it ruptures and the bubble either collapses, or merges with neighboring bubbles, and the formation of larger bubbles is observed (7). Magnetic resonance imaging of drainage in a foam and the changes in water proton mobility offer new information towards understanding foam drainage. Since the procedure is non-invasive and non-destructive, dynamic processes may be observed on a single sample, without disruption, for the duration of the entire experiment.

Signal intensity is proportional to proton density and the gravimetric distribution of water in the draining egg white foam. Distribution of density as function of time and position is shown in Figure 5. For approximately 28 min, there is an increase in density near the screen, at the bottom of the sample. Accumulation to what appears to be a critical mass is followed by separation of the liquid, after which, the density of the foam became relatively constant. At the top, the density in the foam decreased and became constant after approx. 70 min of draining.

Distribution of spin-spin relaxation times as a function of density is shown for a range of positions in Figure 6. The $T_2$' values, i.e., proton mobility, increased at the bottom of the foam sample for the duration of the experiment. One would expect this since at the bottom, the sample became more liquid with time, thus the mobility of protons was expected to approach that in liquid egg whites. With constant draining, however, one would expect that the top of the foam should become drier and more solid-like, restricting mobility, i.e., causing a decrease in the $T_2$' values. However, except near the screen region of the draining sample, the observed trends in $T_2$' were displaying characteristics of the two-state fast exchange behavior.

## TWO-STATE FAST EXCHANGE THEORY

Classically, water in a complex system is characterized to be in either of two states; "bound" or "free". However, this is an over-simplified view, since the protons of water are more appropriately described as ranging from a state of complete "mobility" (e.g., pure water) to a state of being "immobilized" (e.g., water of hydration). Mobility of water is material dependent. Single relaxation time is typical of single phase systems and of multi-phase systems with fast exchange, where the observed relaxation time is a weighted average of the component relaxation times (8). The slopes of relaxation curves correlate to the mobility of water protons in the material, i. e., the steeper the slope, the more restricted the water proton mobility. While spin-lattice relaxation ($T_1$) is influenced only by rapid dynamic processes at about the Larmor frequency, spin-spin relaxation ($T_2$) is affected by both fast and slow molecular processes.

In draining liquid of the egg white foam, which was 87.5% water, only the observed $T_2$ values indicated any molecular activity that would invite further discussion.

There are at least two relaxation mechanism in a complex system which can detract significantly from the magnitude of an observed $T_2$. They are cross-relaxation and proton exchange. Cross-relaxation results from a dipole-dipole interaction between the protons of a macromolecule, such as the proteins in the egg whites, and the protons of water. That is, on coupling of the dipole moments, the water proton transfers its magnetization to the proton of the protein (macromolecule). This makes the relaxation rates of the involved protons equal and the observed $T_2$ shorter, implying that water is less mobile than it actually is. Cross-relaxation does not involve exchange of protons, whereas proton exchange, or "chemical exchange", is a physical exchange of protons between water molecules in distinctly mobile and relatively immobilized states. The magnitude of influence proton exchange has on relaxation rate will depend on the rate of exchange, on the fraction of protons in the two opposite states of mobility , and on the intrinsic relaxation rates of the protons in those states (9). The observed relaxation rate (Equation 4) is a weighted average of the respective relaxation rates:

$$\frac{1}{T_2'} = P_R\left(\frac{1}{T_{2R}} - \frac{1}{T_{2M}}\right) + \frac{1}{T_{2M}} \qquad (4)$$

The term $P_R$ is the probability of water being in a state of restricted mobility, and $T_{2R}$ and $T_{2M}$ are the respective transverse relaxation times for the water in the restricted and mobile states. This two-state fast exchange model predicts a linear relationship between the observed relaxation rate and the probability of water being in a state of restricted mobility.

$P_R$ can relate to concentration, water activity or MRI signal intensity, as illustrated in Figure 7. The curve which departs from linearity represents systems in which hydration number and relaxation rates changed, and where charge repulsion or charge fluctuations are factors to be considered (4).

The trends observed in the relationship of water proton mobility and density in the draining egg whites are shown in Figure 8. Observations near the screen where relaxation rates decreased as proton density increased show the expected trend for a sample which became more liquid with time of draining. In comparison, the curve for observations 30 mm above the screen was also in agreement with the two-state fast exchange model (Figure 7). In general, mobility as a function of density in the egg white foam was consistent with the two-state fast exchange model in most of the draining sample. Figures 9 and 10 demonstrate changes in relaxation rate at the given position as a function of density in the draining foam. In Figure 10, following the time arrows, a steady accumulation between 0 and 28.6 min was followed by a steady loss of density. Relaxation rates decreased for the duration of the experiment, first indicating an increase in mobility as density increased, then, a decrease as the bubble size increased further reducing the fraction of protons with restricted mobility.

Measurement of foam collapse and/or of bubble size was beyond the scope of this work. However, this type of information may be gained, also, when considering water proton mobility. An example is given in Figure 11. The values of $T_2'$ are plotted as a function of density five mm from the top of the foam. Between 44 and 70.4 min, a shift in density is observed. Since nothing was added to the system, it can only be due to the collapse of some of the bubbles above, supplying a limited flow of the liquid released by the collapse or merging of bubbles. By observation, at the end of the experiment, the foam volume changed very little, but the bubbles increased in both size and fragility near the top of the sample.

In conclusion, the procedure for measuring the positional mobility of water in the draining egg white foam has been developed and its use demonstrated. It is important to set up the experiments where they yield good signal to noise ratio to reduce the effect of noise to minimal level for estimation of $T_2$ values. For this same reason, the edges of projections (extreme top and bottom) were also omitted from the calculations of $T_2$. The study showed that the spin-spin relaxation time varied in the foam as a function of density and bubble size.

ACKNOWLEDGEMENTS

The authors thank Henningsen Foods, Inc. for supplying the dehydrated egg whites, and the University of California, Davis, NMR Facility for partial support of this study.

LITERATURE CITED

1. Morris P. G., Nuclear Magnetic Resonance Imaging in Medicine and Biology, Clarendon Press, Oxford University Press, New York (1986).

2. Weisser H., "NMR spectroscopy in the food industry," Minispec NMR Spectroscopy Application Note, IBM Instruments Inc. 12/84, (1984).

3. Lillford P. J., A. H. Clark and D. V. Jones., "Distribution of water in heterogeneous food and model systems," In Water in Polymers, M. J. Comstock (ed)., ACS Symposium Series

127. American Chemical Society, Washington, D.C. (1980).

4. Richardson S. J. and M. P. Steinberg., "Applications of nuclear magnetic resonance," In Water Activity: Theory and Applications to Food, L. B. Rockland and L. R. Beuchat (eds). Marcell Dekker, Inc., New York (1987).

5. Farrar, T. C., An Introduction to Pulse NMR Spectroscopy, Farragut Press, Chicago (1987).

6. Leonard, R. A. and Lemlich, R., "A study of interstitial liquid flow in foam," J. A. I. Ch. E., 11, 18 (1965).

7. Barber, A. D. and Hartland, S. "The collapse of cellular foams," Trans. Instn Chem. Engrs, 53, 106 (1975).

8. Leung, H. K., Steinberg, M. P., Wei, L. S. and Nelson, A. I., "Water binding of macromolecules determined by pulsed NMR," J. Food Sci., 41, 297 (1976).

9. Berliner, L. J. and Reuben, J., Biological Magnetic Resonance, Plenum Press, New York (1980).

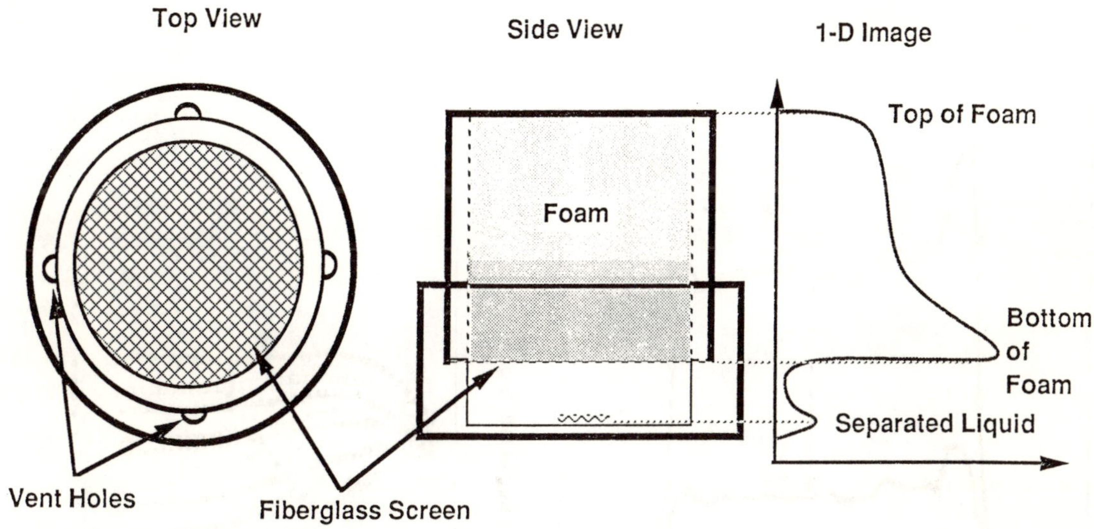


Figure 1. Schematic illustration of the 5 * 5 cm round sample holder and a representative 1-D image. Vent holes were needed to allow the escape of air when the foam column was inserted into the drip catcher. The 1-D image attempts to illustrate that signal intensity is proportional to the numbers of spin active nuclei in a corresponding slice (disk) of sample in the round container.

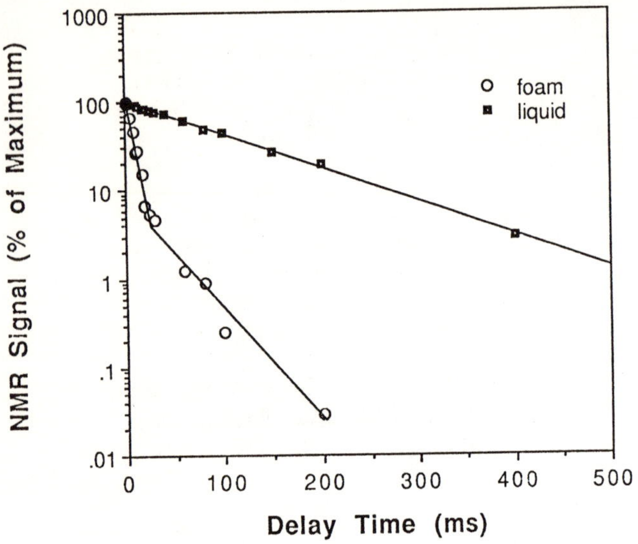


Figure 2. Typical spin-spin relaxation curves obtained by the $\pi/2 - \tau - \pi - \tau$ pulse sequence in liquid and foamed egg whites. The curve for the egg white foam implies that water protons are in multiple states of mobility.

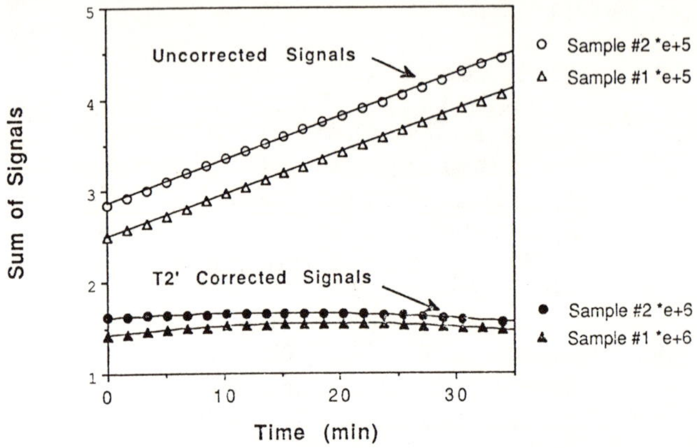


Figure 4. Comparison of sums of uncorrected signals and of signals which were positionally $T_2'$ corrected. Sums of uncorrected signals indicated that spin-spin relaxation time (proton mobility) in the foam varied as a function of proton density. In comparison, the sums of the $T_2'$ corrected signals were significantly higher than the uncorrected sums, and remained relatively constant, as one would expect after proper correction for variable proton mobility.

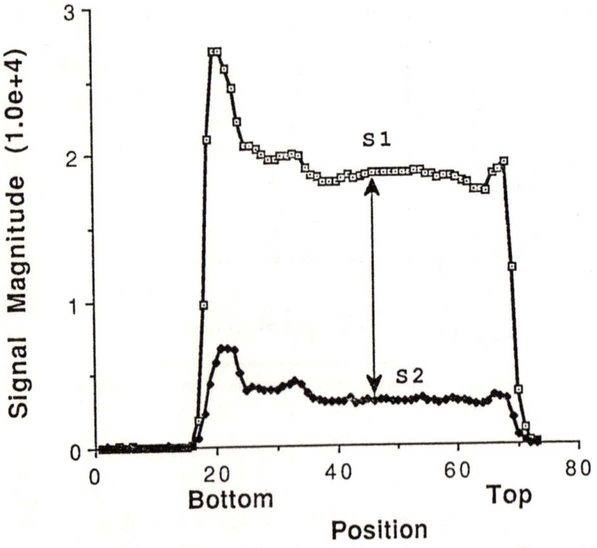


Figure 3. Comparison of 1-D images obtained with 10 ms ($S_1$) and 20 ms ($S_2$) echo delay times. For calculations of $T_2'$, regions expected to be sources of significant error, such as the noise in the baseline or the rapid changes seen at the top and bottom edges, were not included.

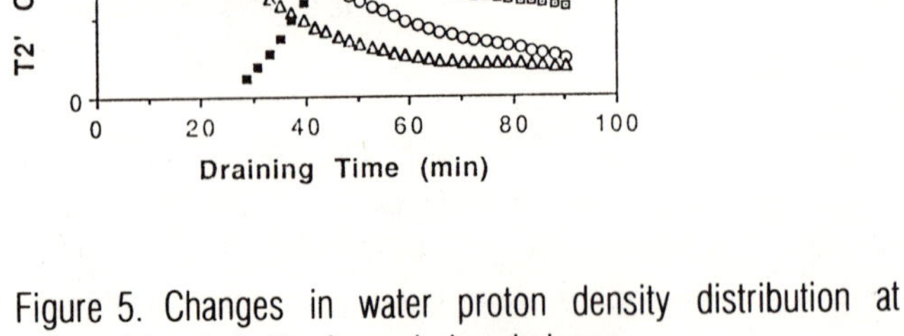


Figure 5. Changes in water proton density distribution at selected levels in the foam during drainage.

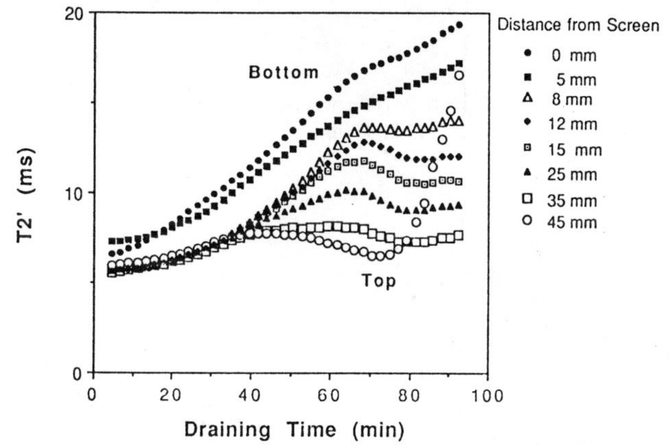


Figure 6. Changes in positional spin-spin relaxations times during foam drainage.

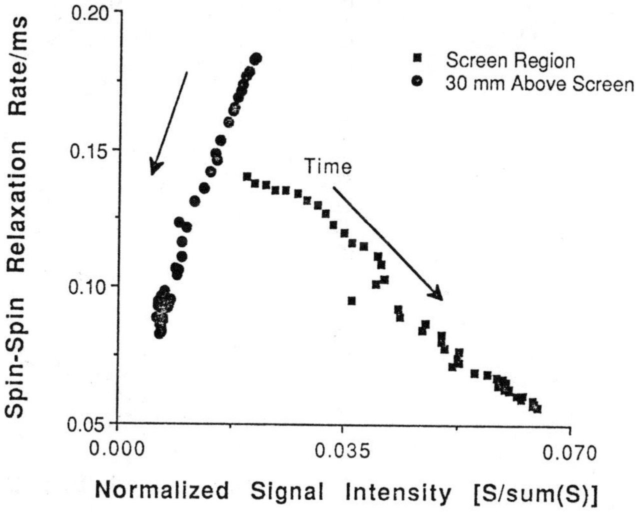


Figure 8. Opposing trends in relaxation rate as function of proton density observed in draining egg whites.

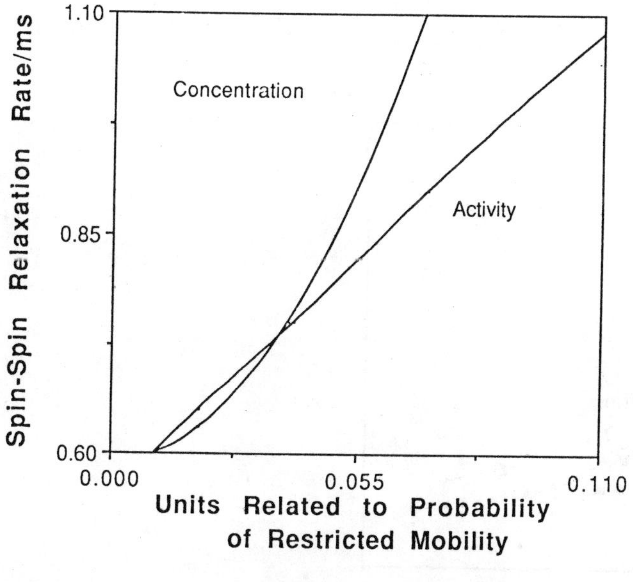


Figure 7. An illustration of the relationship between the observed relaxation rate and the probability that water protons are in a state of restricted mobility, as predicted by the two-state fast exchange model. Departure from linearity can be caused by changing hydration numbers and/or relaxation rates, and/or by charge repulsion/fluctuations.

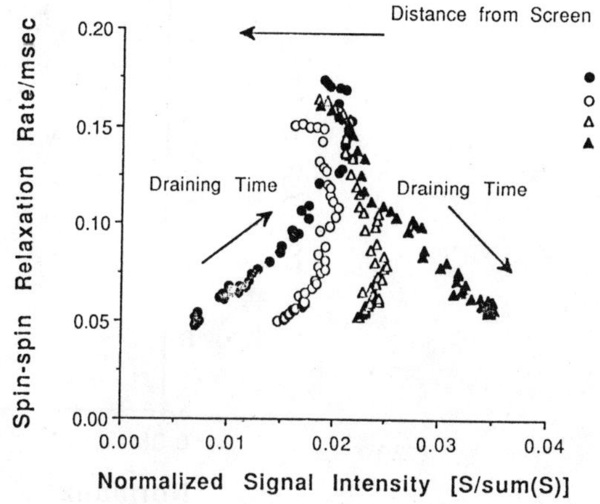


Figure 9. Changing of spin-spin relaxation rate as function of proton density (time and position) in a region between 5 mm and 15 mm from the screen.

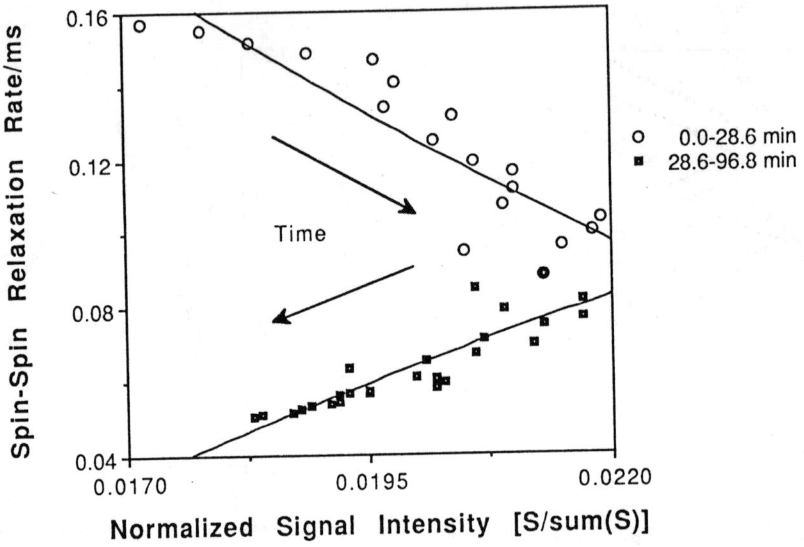


Figure 10. Changing of spin-spin relaxation rate as function of proton density at ten mm above the screen.

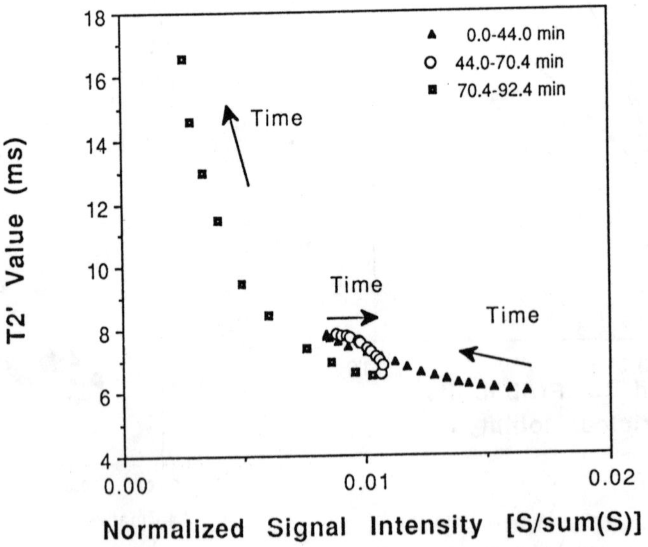


Figure 11. Spin-spin relaxation time as function of density (signal intensity) five mm from top of foam, showing a period of changing bubbles.

# INDEX

# SYMPOSIUM SERIES

## ADSORPTION

96 Developments in Physical Adsorption
117 Adsorption Technology
219 Recent Advances in Adsorption and Ion Exchange
230 Adsorption and Ion Exchange—'83
233 Adsorption and Ion Exchange—Progress and Future Prospects
242 Adsorption and Ion Exchange: Recent Developments
264 Adsorption and Ion Exchange: Fundaments and Applications

## AEROSPACE

33 Rocket and Missile Technology
52 Chemical Engineering Techniques in Aerospace

## BIOENGINEERING

69 Bioengineering and Food Processing
84 The Artificial Kidney
86 Bioengineering ... Food
93 Engineering of Unconventional Protein Production
99 Mass Transfer in Biological Systems
108 Food and Bioengineering—Fundamental and Industrial Aspects
114 Advances in Bioengineering
163 Water Removal Processes: Drying and Concentration of Foods and Other Materials
172 Food, pharmaceutical and bioengineering—1976/77
181 Biochemical Engineering Renewable Sources of Energy and Chemical Foodstocks

## CRYOGENICS

224 Cryogenic Processes and Equipment 1982
251 Cryogenic Properties, Processes and Applications 1986

## CRYSTALLIZATION

110 Factors Influencing Size Distribution
193 Design Control and Analysis of Crystallization Processes
215 Nucleation, Growth and Impurity Effects in Crystallization Process Engineering
240 Advances in Crystallization From Solutions
253 Fundametnal Aspects of Crystallization and Precipitation Processes

## DRAG REDUCTION

11 Drag Reduction
130 Drag Reduction in Polymer Solutions

## ENERGY

### Conversion and Transfer

5 Heat Transfer, Atlantic City
57 Heat Transfer, Boston
59 Heat Transfer, Cleveland
75 Energy Conversion Systems
79 Heat Transfer with Phase Change
87 Advances in Cryogenic Heat Transfer
113 Convective and Interfacial Heat Transfer
118 Heat Transfer—Tulsa
119 Commercial Power Generation
138 Heat Transfer—Research and Design
162 Energy and Resource Recover from Industrial and Municipal Solid Wastes
174 Heat Transfer, Research and Application
189 Heat Transfer—San Diego 1979
202 Transport with Chemical Reactions
208 Heat Transfer—Milwaukee 1981
216 Processing of Energy and Metallic Minerals
225 Heat Transfer—Seattle 1983
236 Heat Transfer—Niagra Falls 1984
245 Heat Transfer—Denver 1985
257 Heat Transfer—Pittsburgh 1987
263 Heat Transfer—Houston 1988
269 Heat Transfer—Philadelphia 1989

### Nuclear Engineering

53 Part XIII
56 Part XIV
95 Part XX
104 Part XXI
106 Part XXII
119 Commercial Power Generation
168 Heat Transfer in Thermonuclear Power Systems
169 Developments in Uranium Enrichment
191 Nuclear Engineering Questions Power Reprocessing, Waste, Decontamination Fusion
221 Recent Developments in Uranium Enrichment

## ENVIRONMENT

78 Water Reuse
97 Water—1969
115 Important Chemical Reactions in Air Pollution Control
122 Chemical Engineering Applications of Solid Waste Treatment
124 Water—1971
126 Air Pollution and its Control
133 Forest Products and the Environment
137 Recent Advances in Air Pollution Control
139 Advances In Processing and Utilization of Forest Products
144 Water—1974: I. Industrial Wastewater Treatment
145 Water—1974: II. Municipal Wastewater Treatment
146 Forest Product Residuals
147 Air: I. Pollution Control and Clean Energy
148 Air: II. Control of $NO_{xx}$ and $SO_x$ Emissions
149 Trace Contaminants in the Environment
151 Water—1975
156 Air Pollution Control and Clean Energy
157 New Horizons for the Chemical Engineer in Pulp and Paper Technology
165 Dispersion and Control of Atmospheric Emissions, New-Energy-Source Pollution Potential
170 Intermaterials Competition in the Management of Shrinking Resources
171 What the Filterman Needs to Know About Filtration
175 Control and Dispersion of Air Pollutants: Emphasis on $NO_x$ and Particulate Emissions
177 Energy and Environmental Concerns in the Forest Products Industry
184 Advances in the Utilization and Processing of Forest Products
188 Control of Emissions from Stationary Combustion Sources Pollutant Detection and Behavior in the Atmosphere
195 The Role of Chemical Engineering in Utilizing the Nation's Forest Resources
196 Implications of the Clean Air Amendments of 1977 and of Energy Considerations for Air Pollution Control
198 Fundamentals and Applications of Solar Energy
200 New Process Alternatives in the Forest Products Industries
201 Emission Control from Stationary Power Sources: Technical, Economic and Environmental Assessments
207 The Use and Processing of Renewable Resources—Chemical Engineering Challenge of the Future
209 Water—1980
210 Fundamentals and Applications of Solar Energy II
211 Research Trends in Air Pollution Control: Scrubbing, Hot Gas Clean-up, Sampling and Analysis
213 Three Mile Island Cleanup
223 Advances in Production of Forest Products
232 Applications of Chemical Engineering in the Forest Products Industry
239 The Impact of Energy and Environmental Concerns on Chemical Engineering in the Forest Products Industry
243 Separation of Heavy Metals and Other Trace Contaminants
246 Advances in Process Analysis and Development in the Forest Products Industries.
265 Resource Recovery of Municipal Solid Wastes

## FLUIDIZATION

101 Fundamental Processes in Fluidized Beds
105 Fluidization Fundamentals and Application
116 Fluidization: Fundamental Studies Solid-Fluid Reactions, and Applications
176 Fluidization Application to Coal Conversion Processes
205 Recent Advances in Fluidization and Fluid-Particle Systems
234 Fluidization and Fluid Particle Systems: Theories and Applications
241 Fluidization and Fluid Particle Systems: Recent Advances
255 New Developments in Fluidization and Fluid-Particle Systems
262 Fluidization Engineering: Fundamentals and Applications
270 Fluidization and Fluid Particle Systems: Fundamentals and Application
276 Advances in Fluidization Engineering

## HISTORY OF CHEMICAL ENGINEERING

100 The History of Penicillin Production
235 Diamond Jubilee Historical/Review Volume

## ION EXCHANGE

79 Adsorption and Ion Exchange Separations
219 Recent Advances in Adsorption and Ion Exchange
230 Adsorption and Ion Exchange—'83
233 Adsorption and Ion Exchange—Progress and Future Prospects
259 Recent Progress in Adsorption and Ion Exchange

## KINETICS

25 Reaction Kinetics and Unit Operations
73 Kinetics and Catalysis

## MINERALS

15 Mineral Engineering Techniques
85 Fossil Hydrocarbon and Mineral Processing
173 Fundamental Aspects of Hydrometallurgical Processes
180 Spinning Wire from Molten Metals
216 Processing of Energy and Metallic Minerals

## PETROCHEMICALS

49 Polymer Processing
127 Declining Domestic Reserves—Effect on Petroleum and Petrochemical Industry
135 The Petroleum/Petrochemical Industry and the Ecological Challenge
142 Optimum Use of World Petroleum
212 Interfacial Phenomena in Enhanced Oil Recovery

## PETROLEUM PROCESSING

103 $C_4$ Hydrocarbon Production and Distribution
127 Declining Domestic Reserves—Effect on Petroleum and Petrochemical Industry
135 The Petroleum/Petrochemical Industry and the Ecological Challenge
155 Oil Shale and Tar Sands
226 Underground Coal Gasification: The State of the Art

## PHASE EQUILIBRIA

2 Pittsburgh and Houston
6 Collected Research Papers
88 Phase Equilibria and Gas Mixtures Properties

## PROCESS DYNAMICS

36 Process Dynamics and Control
46 Process Systems Engineering
55 Process Control and Applied Mathematics
214 Selected Topics on Computer-Aided Process Design and Analysis
267 Process Sensing and Diagnostics

## SEPARATION

120 Recent Advances in Separation Techniques
192 Recent Advances in Separation Techniques—II
250 Recent Advances in Separation Techniques—III

## SONICS

109 Sonochemical Engineering

## MISCELLANEOUS

48 Chemical Engineering Reviews
70 Small-Scale Equipment for Chemical Engineering Laboratories
112 Engineering, Chemistry, and Use of Plasma Reactors
125 Vacuum Technology at Low Temperatures
143 Standardization of Catalyst Test Methods
182 Biorheology
183 The Modern Undergraduate Laboratory Innovative Techniques
185 Electro Organic Synthesis Technology
186 Plasma Chemical Processing
187 Chronic Replacement of Kidney Function
194 Hazardous Chemical—Spills and Waterborne Transportation
203 A Review of AIChE's Design Institute for Physical Property Data (DIPPR) and Worldwide Affiliated Activities
204 Tutorial Lectures in Electrochemical Engineering and Technology
206 Controlled Release Systems
217 New Composite Materials and Technology
220 Uncertainty Analysis for Engineers
228 Problem Solving
229 Tutorial Lectures in Electrochemical Engineering and Technology—II
231 Data Base Implementation and Application
237 Awareness of Information Sources
238 New Developments in Liquid-Liquid Extractors: Selected Papers From ISEC '83
244 Experimental Results from the Design Institute for Physical Property Data. I: Phase Equilibria
247 Chemical Engineering Data Sources
248 Industrial Membrane Processes
249 Measurement of High Temperatures in Furnaces and Processes
252 Thin Liquid Film Phenomena
254 Electrochemical Engineering Applications
256 Experimental Results From the Design Institute for Physical Property Data: Phase Equilibria and Pure Component Properties
258 Fiber Optics: Processing and Applications
261 New Membrane Materials and Processes for Separation
266 Diffusion and Convection in Porous Catalysts
268 Membrane Reactor Technology
271 Experimental Results From the Design Institute for Physical Property Data: Phase Equilibria and Pure Component Properties—Part II.
272 Membrane Separations in Chemical Industries
273 Fundamentals of Resid Upgrading
274 Competitiveness of the U.S. Chemical Industry in International Markets
275 Design Institute for Physical Property Data: Ten Years of Accomplishment
277 Food Emulsion and Foams: Theory and Practice

## MONOGRAPH SERIES

3 The Manufacture of Nitric Acid by the Oxidation of Ammonia—The DuPont Pressure Process by Thomas H. Chilton
4 Experiences and Experiments with Process Dynamics by Joel O. Hougen
5 Present, Past, and Future Property Estimation Techniques by Robert C. Reid
6 Catalysts and Reactors by James Wei
7 The 'Calculated' Loss-of-Coolant Accident by L.J. Ybarrondo, C.W. Solbrig, H.S. Isbin
8 Understanding and Conceiving Chemical Process by C. Judson King
9 Ecosystem Technology: Theory and Practice by Aaron J. Teller
10 Fundamentals of Fire and Explosion by Daniel R. Stull
11 Lumps, Models and Kinetics in Practice by Vern W. Weekman, Jr.
12 Lectures in Atmospheric Chemistry by John H. Seinfeld
13 Advanced Process Engineering by James R. Fair
14 Synfuels from Coal by Bernard S. Lee
15 Computer Modeling of Chemical Processes by J.D. Seader
16 "High-Tech" Materials by Sheldon Isakoff
17 Separations: New Directions for an Old Field
128 Biotechnology: Status and Perspectives